DATE		

ADSORPTION OF POLYMERS

Yu. S. Lipatov and L. M. Sergeeva

ADSORPTION OF POLYMERS

Translated from Russian by R. Kondor
Translation edited by D. Slutzkin

A HALSTED PRESS BOOK

JOHN WILEY & SONS
New York · Toronto

ISRAEL PROGRAM FOR SCIENTIFIC TRANSLATIONS
Jerusalem · London

© 1974 Keter Publishing House Jerusalem Ltd.

Sole distributors for the Western Hemisphere

HALSTED PRESS, a division of
JOHN WILEY & SONS, INC., NEW YORK

Library of Congress Cataloging in Publication Data

Lipatov, Ĩuriĭ Sergeevich.
  Adsorption of polymers.

  Translation of *Adsorbtsiĩa polimerov.*
  "A Halsted Press book."
  1. Surface chemistry. 2. Adsorption. 3. Polymers
and polymerization. I. Sergeeva, Lĩudmila Mikaĭlovna.
II. Title.
QD506.L5613   547'.84   74-12194
ISBN 0-470-54040-0

Distributors for the U.K., Europe, Africa and
the Middle East

JOHN WILEY & SONS, LTD., CHICHESTER

Distributors for Japan, Southeast Asia and India
TOPPAN COMPANY, LTD., Tokyo and Singapore

Distributed in the rest of the world by

KETER PUBLISHING HOUSE JERUSALEM LTD
ISBN 0 7069.1434.3
IPST cat.no.22116

This book is a translation from Russian of
ADSORBTSIYA POLIMEROV
Izdatel'stvo "Naukova Dumka"
Kiev, 1972

Printed and bound by Keterpress Enterprises, Jerusalem
Printed in Israel

CONTENTS

FOREWORD

One of the most important fields of the physical chemistry of polymers and colloid chemistry is the physical chemistry of surface phenomena in polymers /1, 2/. This is because the creation of new polymeric materials, from polymers used in everyday life to those employed in space engineering, involves the application of heterogeneous polymeric systems. It is clear that most modern polymeric materials are heterogeneous systems, with highly developed interfacial areas. These polymers include reinforced plastics, filled thermoplastics, reinforced rubber, lacquer-varnish coatings, glues, and others.

Therefore, surface phenomena in polymers and polymeric materials are very important for all their properties. This applies, in particular, to the structural and mechanical properties. Thus, a study of the behavior of macromolecules on the interface is one of the most important tasks in this field. When we speak of surface phenomena in polymers, we should note that they are important not only from the technical point of view, but also from the biological one, since the role of surface phenomena in which the molecules of biopolymers participate is also very significant. Finally, surface phenomena are also important for solving problems of a newly developing field, namely, the application of polymers to medicine. Here surface phenomena occur on the interface with living tissue.

The problem of surface phenomena in polymers has many aspects. It includes questions such as the adhesion of polymers to solid surfaces, the structure and properties of monolayers, the structural and mechanical properties of the boundary layers of polymers in contact with solid bodies, etc. However, all these questions are closely related to a central problem, namely, the adsorption of polymers on solid surfaces.

Adhesional interaction on a polymer-solid body interface is mainly adsorptional interaction between two bodies. The adsorption of polymers on the surface of a solid body determines the structure of the boundary layer, the type of packing of the macromolecules in the boundary layer, and hence the molecular mobility of the chains, and their relaxation and other properties. Adsorptional processes play a role not only in the complex of the final physicochemical and physicomechanical properties of the polymeric macromolecule, its processing, or synthesis, when these processes take place in the presence of solid bodies of another character, such as fillers, pigments on the surface of metals, glass, etc. The formation of glued joints, the application of varnish-lacquer coatings, and many other technological processes, include the adsorption of polymers on the surface as the first stage. This shows the importance of studying the adsorption processes of polymers on a solid surface in most technological processes.

We shall deal in more detail with the importance of adsorption in the complex of problems of surface phenomena.

In spite of the fact that adhesion processes are discussed in world and Soviet literature in a large number of papers /3—14/, the true molecular mechanism of adhesion is still considered to be nuclear. The existing and developing theories describe partial and limited phenomena. The electrical adhesion theory /3, 4/ considers electrical phenomena that arise when the adhesive is removed from the support, but does not and cannot explain adhesion itself, because electrical phenomena arise during separation, and we are interested in the case when the adhesion bond is unimpaired. The diffusional theory of adhesion /14/ can in practice be applied to the mutual adhesion of polymers only. Thus, the only theory at present applicable is the adsorptional theory of adhesion, which correlates adhesion with the action of intermolecular forces on the interface, that is, with adsorption. Though it has some limitations (as has any theory), from the physical point of view the adsorptional theory is the best. In particular, the concepts on the electrical double layer arising when heterogeneous surfaces are in contact also include adsorption and the orientation of the polar macromolecular groups, that is, these ideas are within the framework of the adsorptional theory /4/. However, the development of this theory has now stopped, because the theory of macromolecular adsorption on solid surfaces has not yet been sufficiently expounded.

In the production of polymeric materials, the processes of wetting solid surfaces by oligomeric or polymeric molecules are very important. Good wetting is a necessary condition for firm adhesion joints and good physico-mechanical properties of the material /15—17/. From the thermodynamic point of view, wetting is the result of the adsorption of polymeric molecules on the interface. Here we have to deal with another very important problem relating to change in the adsorptional interaction on the interface during polymerization or polycondensation, that is, when the low-molecular weight compound is converted into a high-molecular weight one /18/. It is clear that the conditions of formation of adsorptional bonds on the interface should be greatly dependent on the properties of the macromolecule: its molecular weight, chemical nature, and the flexibility of the chain. Thus, it is very important to study the adsorption of polymers as a function of their molecular weight.

The interaction of polymeric molecules with solid bodies appreciably changes all the properties of the polymers because adsorptional interaction on the interface reduces the molecular mobility of the chains during molding and operation. This leads to changes in the boundary layer, in the temperatures at which thermodynamic and structural changes proceed in these layers, and in some other related phenomena /18—21/. However, the structure of the boundary layer and the conditions of its formation depend mainly on the character of the adsorption, and are determined by the structure of the adsorption layer. Thus, intermolecular interactions in filled and reinforced systems also involve adsorption. We shall mention another aspect, namely, the effect of adsorption on the process of synthesizing high-molecular weight compounds on the interface with solid bodies /1/. The adsorption of growing polymeric chains of variable molecular weight and with varying molecular weight distribution appreciably changes the kinetic reaction conditions, and if three-dimensional space networks are required, it affects their structure /22,23/. Hence, adsorptional phenomena are important not only in the processing or application of polymers but also in their synthesis.

The structure of adsorption and boundary layers on the interface with solid bodies is closely related to the surface energy of polymers /15—17, 24—28/. Molecular packing in the adsorption layer and the layer structure can be evaluated from data on monolayers on liquid surfaces. From studies on the special features of the structure of polymeric monolayers it is possible to examine the behavior of boundary layers on polymeric surfaces and not on liquid, and thus to obtain an approximate evaluation of the behavior of polymeric surface layers /29—34/.

An important aspect of surface phenomena in polymers is the role of supermolecular structures in the properties of adsorption layers and monolayers, and in adsorption. We know that overlapping of molecular formations in solutions, that is, the creation of supermolecular structures, begins even in dilute solutions /35/. It is also possible that molecular aggregates are formed in monolayers /36, 31/. The formation of molecular aggregates also appreciably affects adsorption /37, 38/.

Processes of the adsorption polymers were first discussed in world literature about 10 to 15 years ago, mainly because of the development of filled and reinforced plastics. Thus, the adsorption of polymers from solutions can be considered to be one of the most important fields in the physical chemistry of polymers /1, 2, 18/. However, so far there has been no correlation of experimental data with theoretical concepts, unless we mention the out-of-date paper of Patat et al. /39/. Such a generalization is very desirable, from the theoretical point of view (the macromolecules may be subjected to considerable changes in the adsorption layer), and to establish general laws describing the adsorption and desorption of polymers on solid bodies, and to explain the ways in which these processes differ from those observed for low-molecular weight compounds /40/.

The generalization and development of ideas on adsorption may be the starting point for further developments in the physical chemistry of filled and reinforced polymers, and also in the physical chemistry of nonwoven polymeric materials which are so important in modern industry /41—44/.

The formulation and solution of these problems is a task which the authors of this book have attempted to fulfill. The basic material presented gives the results of research on polymer adsorption and related problems published in Russian and non-Russian literature, and the results of the authors' own research and that of their co-workers. We have tried to present general ideas on adsorption and reveal general adsorption patterns. Thus, we do not claim to have given a complete review of all the papers available in the literature on the problems in question, as this would have diverted us from our appointed task and greatly increased the size of the monograph.

Chapters I, IV—VII, and the conclusion were written by Yu.S. Lipatov, and Chapters II and III by L. M. Sergeeva.

Chapter 1

METHODS FOR STUDYING ADSORPTION

DETERMINATION OF THE ADSORPTION OF POLYMERS FROM SOLUTIONS

Various methods are being used for studying adsorption kinetics and equilibrium adsorption. Most of these methods are similar to the ones employed for studying the adsorption of low-molecular weight substances. The simplest and most popular method is that of mixing a given weight of the adsorbent with a given volume of polymer solution of known concentration. The adsorption system is immersed into a thermostat and held until equilibrium is established. The adsorption is determined from the changes in the concentration of the solution. In this case, either the content of the polymer is varied while the amounts of solvent and adsorbent are kept constant, or the amount of agent is varied and the amounts of solvent and polymer are kept constant. The concentrations of the solutions are measured by gravimetry, nephelometry, infrared and ultraviolet spectroscopy, viscometry, the method of labeled atoms, etc.

Thus, when the adsorption of polyesters was studied on silica, alumina, and glass, the concentration of the polymer solutions was determined from the carbonyl band with $\lambda_{max} = 5.8 \, \mu$ on a two-beam infrared spectrophotometer /45/. The error of the determination was 1 mg/ml. The concentration of poly(vinyl acetate) and poly(vinyl acrylates) was determined from the absorption band of the carbonyl group with $\lambda_{max} = 1740$ cm^{-1} /46/. (The spectra were taken on the UR-10 spectrophotometer.)

The spectral method for studying the adsorption of poly(vinyl acetate) on glass powder in various solvents /47/ showed that the characteristic band of the carbonyl group shifts, depending on the nature of the solvent. For example, for benzene the band appears at 1740, for toluene at 1738, for chloroform at 1734, and for carbon tetrachloride at 1741 cm^{-1}.

Sometimes radioactive polymers are used in adsorption studies /48–51/. Thus, in studies on the adsorption of linear polydimethylsiloxane /50/ on different surfaces (carbon black, TiO_2, Fe_2O_3, and others), radioactive ^{31}Si was used.

Tritium-labeled styrene which was polymerized by the anionic mechanism was used for studying the adsorption of polystyrene from solutions in cyclohexane on a chromium surface /49, 50/. The concentration of the adsorbed polymer was determined by a Geiger counter with automatic recording. With this method very dilute solutions (10^{-1} to 10^{-4} mg/ml) can be determined with a high accuracy.

The adsorption can also be determined chromatographically /51—53/. The adsorbent is placed into an adsorption column with a glass filter.

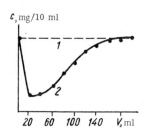

c, mg/10 ml

20   60   100   140 V, ml

FIGURE 1. Auxiliary curves for determining adsorption by chromatography:

1) initial polymer solution;
2) solution after passage through column with pigment.

The polymer solution emerging from the column is collected in several fractions, and the polymer content in these fractions is determined by conventional methods. To find the total amount of polymer adsorbed at the given concentration, an auxiliary curve is plotted. The polymer content in the fractions collected is plotted on the ordinate, and the amount of solution that has passed through is plotted on the abscissa (Figure 1). From this auxiliary curve we plot the kinetic adsorption curve for each concentration. The adsorption itself is determined from the overall amount of adsorbed polymer, which is equal to the area between curves 1 and 2. The experimental error is ± 3% /52/.

This method is very suitable when studying stable suspensions of highly dispersed absorbents in polymer solutions (in this case it is difficult to separate equilibrium solutions from the adsorbent by centrifugation).

In the chromatographic method, knowledge of the adsorption isotherm and of its dependence on the molecular weight is very important. Therefore, we must separately determine the adsorption of the polymer on the column packing, or calibrate the column by monodispersed substances to refer the concentration obtained to a given molecular weight. Moreover, the adsorption — concentration function must be linear, and this is achieved at very low concentrations only. With high molecular weights, no equilibrium is established in the column, and kinetic factors affect the chromatographic separation.

However, comparable data on the polydispersity of the samples can be obtained by chromatography without preliminary calibration. The application of adsorption chromatography opens up wide possibilities for studying macromolecules /211/.

All these methods impair the adsorption equilibrium if the adsorbent is separated from the solution to determine concentrational variations. From this point of view the method of Patat and Schliebener is interesting /88/. With this method it is possible to study adsorption without impairment of the adsorption equilibrium, and to directly observe the establishment of equilibrium during the adsorption process.

The main part of the setup is a sensitive balance, with which the concentrational variation in the polymeric adsorbent immersed in the solution is determined. The adsorption system is hermetically sealed to prevent losses of the solvent during the determination, and thus to avoid errors when determining the concentration of the solution from which the polymer is sorbed. The weight variation ΔP measured is made up of the weight increment ΔG due to adsorption minus the buoyancy (lifting force) ΔA. If the initial weight of the adsorbent is P_0 and its weight after adsorption is P, we have

$$P_0 = G_{fo} - A_{fo} + G_{pol} - A_{pol}, \qquad (1.1)$$
$$P = G_{fo} - A_{fo} + G_{res.sol} - A_{res.sol} + G_{pol} - A_{pol}.$$

Then

$$\Delta P = G_{pol} - A_{pol} + \Delta L, \qquad (1.2)$$

where G and A are the weight of the adsorbent foil (fo), polymer (pol), and solvent (sol) in air and in liquid, respectively.

The variation in weight ΔL due to the solvent is very small compared with the other magnitudes. By expressing the buoyancy in terms of the specific gravity we can write

$$G_{pol} = m = \frac{\Delta P}{1 - \dfrac{d_{sol}}{d_{pol}}}. \qquad (1.3)$$

The reproducibility of the method is within $\pm 5\%$. Because the method is very sensitive it can be employed for studying adsorption on smooth surfaces.

The adsorption of polymers from solutions on smooth surfaces (especially glass) can be determined by means of conventional capillary viscometers. If the polymer solution flows through the capillary viscometer the concentration is changed as the result of adsorption of the polymer on the viscometer walls /66, 67/. From the experimental dependence of the viscosity of the solution on its concentration when there is adsorption, we can calculate the amount of adsorbed polymer, and if we know its density, we can calculate the thickness of the adsorption layer.

When we determine adsorption in capillary viscometers we must bear in mind that we find the thickness of the adsorption layer under hydro-dynamic conditions which must correspond to the thickness of the layer under steady-state equilibrium in the quiescent state. Moreover, the density of the adsorption layer cannot be experimentally determined, and the theoretical value may vary over a wide range.

All the known methods have several drawbacks. Some of them are due to the basic nature of polymer adsorption, for example, the effect of the molecular-weight distribution on the preferential adsorption of high- or low-molecular weight fractions at various stages. Other drawbacks are inherent in the methods themselves. For example, shaking or mixing the solutions to ensure better contact between adsorbent and solution may lead to mechanical degradation of the polymer because of the friction between the adsorbent particles. The adsorption for polar and polarizable mole-cules may be increased because of the electrostatic charge on the particles of the absorbent produced by interaction between adsorbent and solvent. Change in the molecular-weight distribution in the solution during adsorption may lead to errors in the determination of the concentration of solutions when methods such as turbidimetry (sensitive to the molecular-weight distribution) are employed.

Another case is the preferential adsorption of the solvent instead of the polymer when the adsorption may be "negative" because of increase in the concentration of the solution. In all cases we must allow for possible adsorption of the solvent on the adsorbent surface, since in practice it is impossible to select a solvent which cannot be adsorbed. Therefore we must bear in mind that two processes may occur simultaneously, viz., adsorption of the solution, and adsorption of the polymer. Since these processes cannot be separated, such competitive adsorption may distort the results, especially if we attempt to calculate the structure of the adsorption layer from the adsorption by using the size of the adsorbent surface as the starting point.

There are other factors besides the potential competitive adsorption of the solution which complicate the calculation. In most papers, methods are used for determining the adsorption parameters which consist, in principle, in the determination of the specific surface by the adsorption of gases or some organic compounds. Moreover, it is assumed that this area is that of the surface on which the polymer is sorbed.

This assumption, however, is insufficiently substantiated. It is evident that the settling area of the molecules of any gas or polymeric segment will differ. The inner surfaces of adsorbents which, in the case of some porous sorbents, are sufficiently large, may be completely accessible to the molecules of gases and inaccessible to polymeric molecules. The inner surfaces may be accessible to fractions of one molecular weight and inaccessible to fractions of another molecular weight, depending on the molecular-weight distribution. The accessibility of pores in the adsorbent depends also on the nature of the solvent used, since the size of molecules in dilute solution depends on the nature of the solvent.

These limitations make it possible to determine the surface of the sorbent accessible to the polymer. Therefore, calculations based on surfaces (except absolutely smooth, nonporous sorbents) cannot be considered reliable. An analogous point of view was expressed by Perkel and Ullman /54/. Therefore, in many papers the calculations are referred to unit weight of adsorbent and not to unit surface. For the same reason the surface area is measured according to the absorption of larger organic molecules, for example, palmitic acid.

If we know the surface area of the adsorbent, we can calculate the amount of polymer (A_m) that can be adsorbed per unit surface with the formation of a closely packed monolayer

$$A_m = 10^3 \rho h S \ (\text{mg}/\text{g}), \qquad (1.4)$$

where ρ is the density of the polymer; S is the specific surface of the adsorbent; and h is the thickness of the polymeric molecule. In the case of the adsorption of dimethylsiloxane, we know the thickness of the monolayer (5.9 A) from data on the properties of monolayers on an aqueous surface. If in our calculation we take the surface determined from palmitic acid, then A_m = 0.20 and 0.32 mg/g for adsorption on glass and iron. These values are several times less than the experimental ones, but if we calculate the data from the area determined by the BET method, the discrepancy becomes several times greater /54/.

From the methodological aspect, the method of turbidity spectra /55-57/ is interesting in adsorption studies. This method can be applied for solutions containing macromolecule aggregates. The method is based on studying the intensity of scattered light as a function of the wavelength $D = D(\lambda)$. This function is found by plotting the experimental values in $\log - \log$ coordinates, $\log D - \log \lambda$, where the slope of the straight line $d \log D / d \log \lambda = n$ is related to the size of the scattering particles by

$$n = f(\alpha), \quad \alpha = \frac{2\pi \, \overline{r_w}}{\lambda'_{av}},$$
(1.5)

while n is independent of the concentration and (for a certain range of sizes) of $m \left(m = \frac{\mu_r}{\mu_\rho} \right)$, that is, the ratio of the refractive indexes of particle and solvent.

From the experimental values of n and calibration curves we find α, and hence the particle radius $\overline{r_w}$:

$$\overline{r_w} = \frac{\lambda'_{av}}{2\pi} \alpha.$$
(1.6)

For very small particles (Rayleigh scattering), n is constant (equal to 4), so that this method cannot be used. To determine particle sizes larger than 200-2500 A, we must take into account the probable value of m.

Particle sizes can be measured over a wide range of concentrations. The minimum concentration limits are given by the sensitivity limit of the instrument used. In principle, there are no limits to the applicability of the method to higher concentrations.

The concentration (number of particles per cm^3) is determined by

$$N = \frac{\tau}{R},$$
(1.7)

where $\tau = \frac{2.3D}{l}$, cm^{-1}; R is the optical cross section, determined graphically from α (or r_w) for given values of m.

The turbidity spectrum method makes it possible to determine the average number and the size of the macromolecular aggregates in the solution. This is very important when studying adsorption from concentrated solutions, where no individual macromolecules, but their aggregates, pass onto the surface of the adsorbent. This method was used to prove the adsorption of macromolecular aggregates /58, 59/.

By these methods we determine the magnitude of adsorption on the surfaces of adsorbents. However, when adsorption processes are studied, the determination of the structure and thickness of the adsorption layer is important.

METHODS FOR STUDYING THE STRUCTURE
OF ADSORBED FILMS

Ellipsometry

The ellipsometric method has recently become popular for studying the thickness of the surface adsorbed film of macromolecules. It is based on the reflection of polarized light from the polymeric film covering the adsorbent surface.

With this method we can measure the thickness and refractive index of films which adhere to a solid surface, and also the optical constants of the surface. The film thickness determined by this method is much smaller than that measured by interferometry. Moreover, with this method it is possible to find directly the characteristics of the adsorbed films when the adsorbent is immersed in the solution. However, the method can be applied only to mirror surfaces that reflect light, and it can be utilized for metallic adsorbents only. Although this method was first developed for studying the structure of nonpolymeric films, it is very promising for polymers.

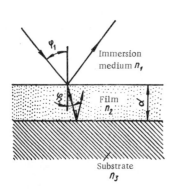

FIGURE 2. Reflection from a film-covered metal surface.

We now explain the principle on which ellipsometry is based. Consider a surface of the adsorbent, with an adsorbed film on it, immersed in a solution (Figure 2) /60, 61/. Such a system can be defined by three refractive indexes, viz., of the sorbent n_3, the film n_2, the solution (solvent) n_1. The beam of polarized light falls on the surface (angle of incidence φ_1), is refracted in the adsorbed film (which is considered to be isotropic), and is reflected from the metallic mirror. For light falling on the liquid-film boundary, the cosine of the angle of refraction is

$$\cos \varphi_2 = \left[1 - \left(\frac{n_1}{n_2} \sin \varphi_1 \right)^2 \right]^{1/2}. \qquad (1.8)$$

Because of reflection from the surface (pure or covered by the adsorbed film), the state of polarization of the reflected light differs from that of the incident light beam. The light components with an electric vector parallel (P) and perpendicular (S) to the mirror plane are reflected in different ways. The parallel and normal coefficients of reflection (Fresnel coefficients) in this case are

$$r_{12}^{P} = \frac{n_2 \cos \varphi_1 - n_1 \cos \varphi_2}{n_2 \cos \varphi_1 + n_1 \cos \varphi_2}, \qquad (1.9)$$

$$r_{12}^{S} = \frac{n_1 \cos \varphi_1 - n_2 \cos \varphi_2}{n_1 \cos \varphi_1 + n_2 \cos \varphi_2}. \qquad (1.10)$$

Accordingly, the reflection coefficients on the film − surface boundary can be expressed in similar terms as r_{23}^P and r_{23}^S. The total reflection coefficient for the two components is then given by Drude's equations

$$R^P = \frac{r_{12}^P + r_{23}^P \exp D}{1 + r_{12}^P \exp D}, \qquad (1.11)$$

$$R^S = \frac{r_n^S + r_{23}^S \exp D}{1 + r_{12}^S r_{23}^S \exp D}. \qquad (1.12)$$

D represents the magnitude

$$D = -4\pi i n_2 \cos \varphi_2 r / \lambda, \qquad (1.13)$$

where λ is the wavelength of the incident light; n_2 is the refractive index of the film; r is the thickness of the film (the required magnitude); $i = \sqrt{-1}$.

The ratio of the normal to the parallel coefficients of reflection is given by

$$\rho = \frac{R^P}{R^S}, \qquad (1.14)$$

or

$$\rho = \tan \Psi \exp (i\Delta), \qquad (1.15)$$

where Δ is the azimuthal angle, and $\tan \Psi$ is the relative phase shift. The magnitude of $\tan \Psi$ is thus a measure of the relative absorption of both components. The other parameters (wavelength, refractive index of the solvent and its concentrational dependence on the dissolution of the polymer, etc.) are determined by additional experiments.

The value of the complex refractive index of a reflecting surface can be determined from the equation

$$n_3 = n_1 \tan \varphi_1 \left[1 - \frac{4\rho \sin^2 \varphi}{(\rho + 1)^2} \right]^{1/2}. \qquad (1.16)$$

The thickness of the adsorbed film is determined as follows. If we know the refractive indexes of the substrate and the film surface then, for a given value of r, we can calculate ρ from (1.8)−(1.14) and $\tan \Psi$ and Δ from (1.15). To illustrate this, the curves calculated for 3 values of n_2 are shown in Figure 3. The numbers along the curves give the thickness of the films in angstroms. The experimental values of $\tan \Psi$ and Δ are then interpolated to determine the thickness of the film.

A more convenient method is the direct solution of an equation with the film thickness as one of its terms. We substitute (1.11)−(1.13) into (1.15) and reduce it to the quadratic form

$$C_1 (\exp D)^2 + C_2 \exp D + C_3 = 0, \qquad (1.17)$$

where C_1, C_2, and C_3 are complex functions of the refractive indexes of angles of incidence Δ and Ψ. For given values of the coefficients we can find two solutions for D, from which we can calculate two values of r. Since these coefficients are complex magnitudes, the thicknesses calculated will also be complex magnitudes. However, the true film thickness must be a real number as it represents a real value.

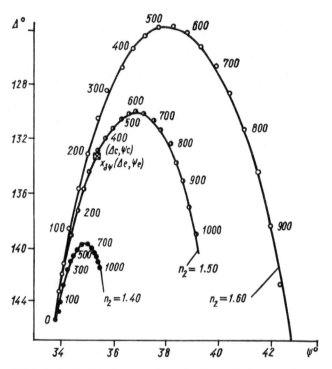

FIGURE 3. Graph of Δ plotted against ψ for films with different refractive indexes in a medium with $n = 1.359$.

These equations cannot be solved if the thickness and n_2 are unknown. For these cases a series of refractive indexes is assumed, and a thickness is computed from experimental values (Figure 3). We shall assume that the experimental points Δ_e and Ψ_e for a film with unknown n_2 are near one of the three curves. In this case we arbitrarily select the refractive index of the film (in Figure 3, $n_2 = 1.5$) and calculate the magnitude corresponding to point Δ_c, Ψ_c with an accuracy also graphically determined, as shown in Figure 3 (errors $\delta\Delta$ and $\delta\Psi$).

Since in this procedure the imaginary components are neglected, these determinations give the thickness with errors that lie within the permissible experimental errors.

The schematic diagram for determining the film thickness is shown in Figure 4. The light from a mercury lamp is filtered, and the 5461 Å line is separated. The monochromatic light passes through a system of lenses in a polarizer, and by means of a birefringent plate, usually a quarter of a wavelength thick, the polarized light is transformed into elliptically polarized light. The latter falls at an azimuthal angle of 70 ±0.01° onto a metallic mirror and is reflected from it. The mirror is immersed into the polymer solution. The reflected light is amplified by an analyzer and photomultiplier and measured by a photometer.

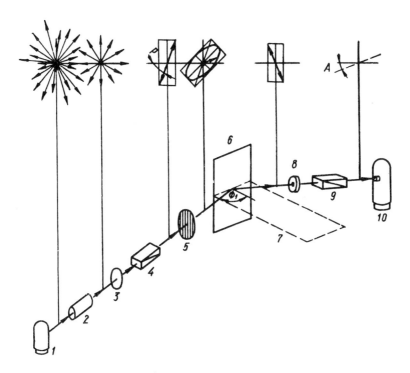

FIGURE 4. Component parts of an ellipsometer:

1) light source; 2) filter; 3) lens; 4) polarizer; 5) compensator (λ/4 plate); 6) reflecting surface (metallic mirror); 7) incident plane; 8) aperture; 9) analyzer; 10) photocell.

The application of this method for calculating the film thickness is based on the assumption that the film is homogeneous and has a discrete boundary and that the concentration of the polymer in it is constant. However, the adsorbed polymer film is not homogeneous and the refractive index is not constant over the film thickness, but changes with the distance from the surface. Therefore, the theory gives only the average "effective" film thickness. The effect of the surface film on the variation in Ψ and Δ depends on the difference between the refractive indexes of the film and the solvent, $n_f - n_s$, and also on the film thickness r. For a nonhomogeneous film, $n_f - n_s$ is a function of the distance from the surface. The thickness r_n and the refractive index n_h of an equivalent homogeneous film are found from

$$\int\limits_0^\infty (n_f - n_s)\, d\,(r) = (n_h - n_s)\, r_n.\qquad(1.18)$$

The value n_h can be found from the equation

$$n_h = \frac{\int\limits_0^\infty n_f\,(n_f - n_s)\, dr}{\int\limits_0^\infty (n_f - n_s)\, dr},\qquad(1.19)$$

which defines the mean refractive index, and in which $n_f - n_s$ acts as the distribution function. From these equations we obtain the equivalent thickness of the homogeneous film

$$r_h = \frac{\left[\int\limits_0^\infty (n_f - n_s)\, dr\right]^2}{\int\limits_0^\infty (n_f - n_s)^2\, dr}.\qquad(1.20)$$

Thus, if the distribution function $(n_f - n_s)$ is known, we can ellipsometrically determine the equivalent magnitudes n_h and r_h. Magnitude r_h is related to the mean square thickness by the equation

$$r_r^2 = \int\limits_0^\infty (n_f - n_s)\, r^2 dt / \int\limits_0^\infty (n_f - n_s) dr.\qquad(1.21)$$

For linear distribution $(n_f - n_s)\, r\, /r_r = 1.75$, for exponential it is 1.47, and for Gaussian it is 1.73.

By the ellipsometric method we can measure the time dependence of parameters $\tan \Psi$ and Δ and thus observe the course of the adsorption process.

Method of attenuated total reflection in the ultraviolet region

To study the structure of the adsorbed film, the attenuated total reflection (ATR) method in ultraviolet light is frequently used /63/. The polymer is adsorbed on a quartz prism in contact with a solution of the polymer. The light transmitted through the prism is totally reflected at the boundary between the prism and the solution if the angle of incidence exceeds some critical value. However, if adsorption takes place at the boundary and the adsorbed film absorbs radiation, some of the incident light is absorbed by the film and attenuated total reflection is

observed; this is a phenomenon well known in spectroscopy. The change
in the reflection pattern depends on the film thickness.

By using ATR we measure the absolute values of two reflection
coefficients, in contrast to the case with ellipsometry in which the ratio
of the reflection coefficients and a phase shift of light are measured.
The sensitivity of the method depends on the differences between the
refractive indexes of the film, the substrate, and the surrounding medium.
The advantage of this method over ellipsometry is that it makes it
possible to evaluate the distribution of the segments in the film if the
incident light does not penetrate to a depth exceeding the film thickness.
Therefore UV light is used, since the depth of penetration decreases with
decrease in wavelength.

The calculation on which the ATR method for the determination of the
adsorbed film is based is to a certain degree similar to the calculation
in ellipsometry.

The ATR method uses two light beams polarized parallel and perpendicular to the plane of incidence. The reflection coefficients are determined
by Fresnel's equations (1.9) and (1.10). The total reflection coefficient R^p
is found by Drude's equations (1.11) and (1.12). The expression (1.13)
for D contains the complex refractive index N of the solution or the
adsorbed film measured by the ATR method and defined as

$$N = n\,(1 - ik),\qquad(1.22)$$

where n is the real part of the refractive index, and k is a magnitude
directly related to the adsorbance A obtained from transmission spectra.
According to Lambert – Beer's law

$$A = \alpha cl/2,3\qquad(1.23)$$

(α is a constant; l is the length of the cell; c is the concentration of
the solution). The value of c for the solution or film measured by the
ATR method is related to α by

$$\alpha c_{ATR} = 4\pi nk/\lambda.\qquad(1.24)$$

In this case k is

$$k = \frac{2.3A}{l}\left(\frac{c_{ATR}}{c}\right)\frac{\lambda}{4\pi n}.\qquad(1.25)$$

After the constants entering Drude's equation have been found we can
calculate the other unknowns (usually using computers).

In addition, we must experimentally determine the ordinary and
extraordinary refractive indexes entering Drude's equation, and this is
extremely difficult. Usually, the reflection coefficients for the given
concentrations of solutions and film thicknesses are determined and the
values obtained are compared with the experimental ones. In the calculations, as in the ellipsometric method, it is assumed that the adsorbed
film is homogeneous and discrete. However, the ATR method can also be

used for calculating the distribution of segments in the adsorbed film. This problem is discussed by Peyser and Stromberg /63/, although no practical studies or theoretical calculations were carried out.

Method for determining the fraction of segments bound to the surface of the adsorbent

To clarify the structure of the adsorbed film and the conformation of macromolecules in this film, it is important to know the fraction of chain segments that interact with the surface since, in contrast to the adsorption of low-molecular weight substances, in the adsorption of polymers all the segments of the polymeric chain can never be simultaneously bound. In this case the magnitude p is important, that is, the ratio of the number of segments attached to 1 cm^2 of surface to the total number of segments in the polymer adsorbed by 1 cm^2. Fontana and Thomas proposed a method for determining the p-value by IR spectroscopy (e. g., OH groups) /64/. Active groups must be on the surface of the polymer, while the polymer segments must contain some other functional group able to form a hydrogen bond with the adsorbent. Thus, when hydrogen bonding occurs between the hydroxyl and carbonyl groups, the absorption band of the carbonyl group ($\nu \sim 1700$ cm^{-1}) is shifted to lower frequencies ($\Delta \nu = 14-29$ cm^{-1}). The segments of the adsorbed macromolecule that are not directly bound to the surface are characterized by the absorption band of the "free" carbonyl groups unperturbed by the hydrogen bond. The number of attached groups can be estimated provided that: 1) the frequencies perturbed and unperturbed by the hydrogen bond can be resolved in the IR spectrum of the adsorbed polymer molecule; 2) an estimate of the extinction coefficients of the adsorbed functional groups can be made.

This method was developed by Peyser et al. /65/, and when experimentally applied it is modified according to the special features of the given system. If the frequencies in the spectra are not satisfactorily separated, differential spectral analysis can be applied (Figure 5). In this case two cells are used. One contains the original solution and the other the polymer adsorbed on the adsorbent, suspended in a solution of the polymer. During adsorption, the adsorption band is shifted toward larger wavelengths because of the reaction between the carbonyl groups of the polymer and the OH groups of the surface (Figure 5).

If we assume that both carbonyl groups of the adsorbed polyester segment are bound to the surface, the fraction of bound segments will depend on the optical density D of the peak in accordance with the equation

$$p = [(D) + \varepsilon_1 c^1]/Lc\,(\varepsilon_1 - \varepsilon_2), \qquad (1.26)$$

where L is the cell length; c is the difference between the initial and final concentrations of the polymer; $\varepsilon_1, \varepsilon_2$ are the extinction coefficients for the free and bound segments at the peak; c^1 is the difference between the concentrations of the polymer solutions in the two cells.

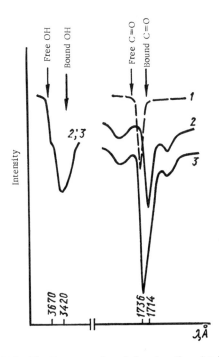

FIGURE 5. The IR spectrum for poly(lauryl methacrylate) in the solution and adsorbed by silica:

1) in dodecane solution, 0.1 wt%; 2) in equilibrium on SiO_2 with the pure solution in the cell; 3) in equilibrium on SiO_2 with poly(lauryl methacrylate) solution in the cell.

It is sufficient to measure D/Lc to correlate changes in p with the final concentration of the polymeric solution after adsorption.

Other methods

There are also other methods for determining the thickness of the adsorbed films. One of them is based on the determination of the sedimentation rate of the adsorbent particles dispersed in the polymeric solution /63/. From the changes in the sedimentation rate in solvent and solution we can find the increase in the particle size, if we assume that it is determined mainly by the formation of an adsorbed film on the surface. This calculation can be carried out if we assume that the adsorbent particles are spherical, and that the adsorbed film has a constant thickness and is impermeable to the solvent.

We know methods based on measurement of the viscosity of polymeric solutions by capillary viscometers /66, 67/. The adsorption of the polymer by the capillary walls affects the viscosity by decreasing the capillary diameter and the concentration of the solution. In the first case the shape of the $\eta_{sp}/c = f(c)$ curve is considerably changed, while in the second case η_{sp}/c decreases. Evidently we can assume that during adsorption on the capillary walls the molecules retain the volume and shape that they have in dilute solutions, if the changes in diameter are sufficient to affect the outflow time.

To calculate the effect of the adsorption layer on η_{sp}/c, let us designate the apparent relative viscosity by $\overset{*}{\eta}_{rel}$, the true relative viscosity by η_{rel}, the capillary radius by r, and the film thickness by a. Then

$$\eta_{rel} = \overset{*}{\eta}_{rel}\left(\frac{a-r}{r}\right)^4 = \eta_{rel}(1 - 4r/a) + \cdots, \qquad (1.27)$$

which yields

$$\overset{*}{\eta}_{sp}/c = \eta_{sp}/c + 4r(\eta_{sp}/c + 1/c)\,1/a. \qquad (1.28)$$

For a given concentration, it can be assumed that a is constant, and in this case there is a linear dependence between $\overset{*}{\eta}_{sp}/c$ and $1/a$.

Thus, if we measure the viscosity with capillaries of various dimensions and plot the $\eta_{sp} = f(a^{-1})$ curve, the intercept on the ordinate gives the η_{sp}/c value, since the adsorbed layer does not affect the outflow time if $a \to \infty$. It was found that the film thickness increased due to adsorption does not change when the capillary is washed by the solvent /66, 67/.

The film thickness can be determined from the variation in the outflow time of the solvent through the capillary according to

$$r = a\Delta t/4t. \qquad (1.29)$$

The viscometric method for determining the thickness of the adsorbed polymer film on the solid particles can be carried out in an entirely different way. This method has the great disadvantage that only the capillary walls are the adsorbents. Ellipsometry and the ATR method are also restricted in their use, since they require mirrorlike smooth surfaces. Therefore, a viscometric method was proposed to study the adsorbed film on the surface of disperse particles. The method is based on the change in volume of particles as the result of the formation of the adsorbed film /68/. The film thickness on disperse particles can be estimated from this. The method is somewhat similar to the sedimentation method, but in contrast to this method it is not restricted by the assumption that the particles are spherical, that the particle size distribution is narrow, etc.

The method is based on determining the volume of the disperse phase by comparing the viscosity of the dispersion and the solution, and applying Einstein's equation. The difference between the true and apparent volumes of the disperse phase is the characteristic effective volume of the adsorbed

film, from which the average thickness of the adsorbed film is calculated.
Einstein's equation for the viscosity of a suspension of rigid spherical
particles in a Newtonian liquid has the form

$$\eta = \eta_0 (1 + K\varphi), \qquad (1.30)$$

where η is the viscosity of the suspension; η_0 is the viscosity of the pure
liquid; φ is the volume fraction of the particles; K is the constant for
spherical particles, equal to 2.5. The formation of the adsorbed film leads
to an increase in the particle size, and therefore constant K has another
value K_{eff}. If we know the diameter of the particles and the values of K
and K_{eff}, we can calculate the thickness of the adsorbed film by equation

$$r = \frac{d}{2} \left(\sqrt[3]{\frac{K_{eff}}{K}} - 1 \right). \qquad (1.31)$$

In the case of a polydisperse adsorbent with a specific surface S_{sp}, we can
determine the effective increase in the volume in the suspension per unit
weight $\Delta\varphi_{eff}$, and thus calculate the thickness of the film

$$r = \frac{\Delta\varphi_{eff}}{S_{sp}}. \qquad (1.32)$$

The special advantage of this method is that it can be applied to
concentrated solutions and melts if the measurements are carried out
under conditions that ensure Newtonian behavior of the system /69/.
Unfortunately, indeterminacies involved in measuring the specific surface
of the sorbed molecules make this method unreliable for such sorbents.
This method can be applied to systems governed by the equation of Einstein.
However, we have many data /70, 71/ indicating that the classical equation
of Einstein is not applicable to many filled systems.

Doroszkowski and Lambourne /68/ applied this method to dilute solutions
in the presence of a small amount of disperse particles (up to 2 vol%),
which do not interact, that is, in the absence of flocculation. To prevent
the flocculation of particles the measurements were carried out over a
sufficiently wide range of shear rates (from 300—4000 sec^{-1}).

The logarithm of the viscosity is plotted against the shear rate $D^{-1/2}$,
and extrapolated to the velocity gradient $D^{-1/2} = 0$. In this way the proposed
method differs from other attempts to estimate the film thickness.

The authors of /68/ assume that particle flocculation may appear at
low shear rates, since at higher rates any structure in the solution
(dispersion) is decomposed, and the particles with an adsorbed film
behave independently. Under these conditions the shape and concentration
of the particles may play a decisive role in the viscosity. By using the
viscosity values obtained by extrapolation to infinite shear rates, we
can estimate the effective hydrodynamic volumes of the particles. The

main difficulty lies in the selection of the most suitable equation, relating the viscosity to the volume of the disperse phase when the system is not described by Einstein's equation. In /68/ an equation of type

$$\eta = \eta_0 (1 + 3x + 23x^2) \qquad (1.33)$$

is used, where x is the volume of the disperse phase. The applicability of this equation to the given systems has been checked.

Chapter 2

THE KINETICS OF ADSORPTION

RATE OF ESTABLISHMENT OF ADSORPTION
EQUILIBRIUM

A study of the kinetics of the establishment of adsorption equilibrium is
of fundamental importance for understanding its mechanism and for
evaluating the reliability of the data obtained on adsorption. The rate
of establishment of equilibrium during adsorption depends on the chemical
nature of the polymer and its molecular weight, the solvent, and the type
of adsorbent. In most cases the kinetic curves indicate an increase in the
amount of adsorbed polymer with time, with an asymptotic approximation
to the equilibrium value (Figure 6).

The diffusion of polymers to the surface of the adsorbent or into its
pores is almost always the stage that determines the adsorption rate. In
the case of porous adsorbents, the equilibrium is rapidly established,
frequently in a few seconds or minutes. Thus, the rate of establishing
equilibrium in the polyester — glass system is 15 sec /73, 120/. In the
system polyester — aluminum, equilibrium is established at a slower rate
(15 min).

For porous adsorbents (e.g., charcoal), equilibrium is established much
more slowly, since diffusion of macromolecules into the pores takes a long
time. The time of establishment of equilibrium in the polystyrene — charcoal
system is 9—10 hr /74/. In other cases, equilibrium is established only
after a few days /75/.

During the adsorption of the copolymer of styrene with butadiene, butyl
rubber, and natural rubber on carbon blacks of various types, equilibrium
was established within 18—90 hours. The time depended on the nature of
the polymer, the specific surface of the carbon black, and other variables.
The highest adsorption rate was noted for butyl rubber, and the lowest for
natural rubber. With increase in the specific surface, the adsorption rate
decreases (see Figure 7) /76/. Sometimes the particles of nonporous
adsorbents, for example, of carbon black, form secondary porous structures,
which are sufficiently stable and are not destroyed when the suspension is
stirred /72/. This secondary porosity can be clearly seen in electron
microscope studies. Porosity hinders the rapid penetration of the macromolecules to the inner surface of the carbon black, and therefore the
establishment of adsorption equilibrium frequently takes several hours
when such adsorbents are used /72/.

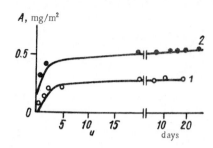

FIGURE 6. Adsorption kinetics of polystyrene with molecular weight 290,000 (1) and 43,000 (2) from CCl$_4$ solutions by carbon blacks.

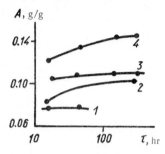

FIGURE 7. Time dependence of the adsorption of the styrene-butadiene copolymer for various specific surfaces of carbon black:

1) 45.6; 2) 114.2; 3) 75.1; 4) 138 m^2/g.

The rate of adsorption of poly(vinyl acetate) on irregularly shaped alumina is much lower than the adsorption rate on the smooth nonporous surfaces of iron or tin /77/. Analogous results were obtained when studying the adsorption of fractionated polystyrene on aluminum and aluminum oxide /78/. The adsorption equilibrium was established within a few hours on the smooth surface of aluminum, and within a few days on porous aluminum oxide.

The adsorption rate on porous adsorbents is determined by the pore size. Thus, in a study of the adsorption of polystyrene from dilute solutions in CCl$_4$ by silica gels /75/, it was found that the adsorption equilibrium on silica gel with a pore diameter of 550 Å and 140−280 Å was attained after a few hours and 30−40 days, respectively. If the dimensions of the macromolecules are larger than the pore size of the adsorbent then the adsorption equilibrium will be established as rapidly as in the case of nonporous substances.

Treatment of the surface of the adsorbent by some substances changes the adsorption kinetics. El'tekov /79/ studied the effect of treatment of the surface of Aerosil silica by silicon-containing substances on the adsorption kinetics of polystyrene dissolved in CCl$_4$. He showed that the degree of covering of the surface of the Aerosil by organic residues was about 75% (analysis for the carbon content). The author believes that incomplete covering of the Aerosil surface by organic groups leads to the formation of a microroughness, which possibly hinders adsorption. The reduction in the adsorption rate by modified Aerosil may be due to this microroughness. Treatment of the Aerosil by boiling water leads to an increase in the particle size and to the appearance of sufficiently wide pores (d = 3000 Å), which also lowers the rate of adsorption.

Stirring during adsorption usually increases the rate at which the system approaches equilibrium. For example, if at a lower stirring rate the adsorption equilibrium in the system polystyrene − poly(methyl methacrylate) — silica gel is established within 0.5—1.5 hr, then when the rate is increased, equilibrium is attained within 3 minutes /80/.

Peterson and Kwei /81/ studied the adsorption kinetics of poly(vinyl acetate) from very dilute benzene solutions (10^{-4}—10^{-6} g/ml) on the surface of chromium plates. It was found that adsorption equilibrium is rapidly established, since the adsorption takes place on a smooth surface from a dilute solution (Figure 8). At a concentration of the solution of $2.3 \cdot 10^{-5}$ mole/liter, the maximum adsorption rate of $1 \cdot 10^{-9}$ mole/cm^3 (curve 1) is reached in a few minutes and then remains constant (Figure 8). At a concentration of less than $5 \cdot 10^{-6}$ mole/liter, the adsorption rate is very dependent on the stirring rate. At a concentration of $4.07 \cdot 10^{-4}$ mole/ /liter (curve 5), the adsorption rate is $1 \cdot 10^{-9}$ mole/cm^2, and is reached within 10 sec; it then increases slowly to $3.8 \cdot 10^{-9}$ mole/cm^2.

The adsorption rate on smooth surfaces is thus seen to increase with decrease in the concentration and on stirring. When adsorption kinetics was studied on disperse adsorbents, Kiselev et al. /46/ noted a decrease in the adsorption rate with decrease in the concentration of the solution.

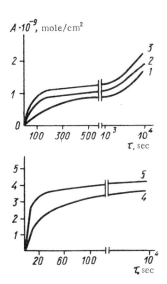

FIGURE 8. Adsorption rate of poly(vinyl acetate) on the surface of chromium at different concentrations of the solutions:

1) $2.3 \cdot 10^{-5}$; 2) $5.75 \cdot 10^{-5}$; 3) $1.15 \cdot 10^{-4}$; 4) $4.07 \cdot 10^{-4}$; 5) $1.27 \cdot 10^{-3}$ mole/liter.

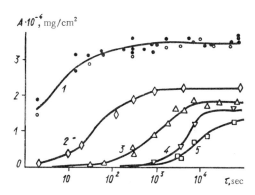

FIGURE 9. Adsorption kinetics of tritium-labeled polystyrene on the surface of chromium at low concentrations of the solutions:

1) concentration of solution 10^{-1}; 2) 10^{-2}; 3) 10^{-3}; 4) $2 \cdot 10^{-4}$; 5) 10^{-4} mg/ml.

The adsorption rate is affected not only by the type of the adsorbent and the concentration of the solution, but also by the molecular weight of the polymer, molecular-weight distribution, temperature, amount of solution, etc. Figure 9 shows the kinetic curves of the adsorption of polystyrene on a chromium surface, and illustrates the dependence of the adsorption rate on the concentration of the solution /49/. In this case the adsorption rate decreases with increase in molecular weight. Thus, for polystyrene with a molecular weight of 38,000, the adsorption peak is attained in 4 hours, while at a molecular weight of 76,000 the adsorption equilibrium is reached only in 24 hours. A similar phenomenon was established for polydimethyl-siloxane, titanium dioxide, etc. (Figure 10) /50/. The adsorption rate of polydimethylsiloxane on silica also depends very strongly on the molecular weight /82/.

Howard and McConnel /84/ studied the adsorption of poly(ethylene oxide) of molecular weights between 390 and 190,000 on activated carbon, Aerosil silica, and nylon. They found that at a molecular weight of 390 equilibrium is established within two hours, but 24 hours are needed for the polymer with a molecular weight of 10,000. This difference is due to the slower diffusion rate of large molecules.

Binford and Gessler /83/ observed a decrease in the adsorption rate of polyisobutylene on carbon black with increase in molecular weight. However, the adsorption rate of poly(vinyl acetate) on activated carbon increases with increase in molecular weight, while the amount of the adsorbed polymer decreases /85/. This unconventional behavior may be due to the fact that with increase in molecular weight the surface accessible to adsorption decreases. Sometimes, however, the adsorption rate is independent of the molecular weight, for example, during the adsorption of polystyrene on silica gel.

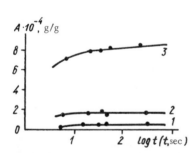

FIGURE 10. Time dependence of the adsorption of polydimethylsiloxane on carbon black:

1) molecular weight 460; 2) 4370; 3) 15,500.

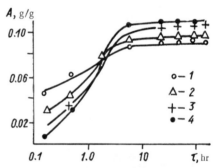

FIGURE 11. Time dependence of the adsorption of polybutadiene of different molecular weights:

1) 10,000; 2) 192,000; 3) 335,000; 4) 502,000.

Bogacheva and El'tekov /72/ proved that the time taken for adsorption equilibrium to be reached is the same for polystyrenes of molecular weights 43,000 and 290,000. Sometimes, for the same polymer both increase and decrease in the adsorption rate with increase in molecular weight are observed. We shall mention the adsorption of polybutadienes with a narrow molecular weight distribution on carbon black HAF from heptane solutions

(Figure 11) /86/. The figure shows that the adsorption rate decreases with increase in molecular weight. The equilibrium adsorption increases so that the curves intersect. Just the opposite occurs with polybutadiene of high molecular weight (Figure 12). The observed decrease in the adsorption rate with increase in molecular weight may be due to the inaccessibility of the adsorption surface to large molecules.

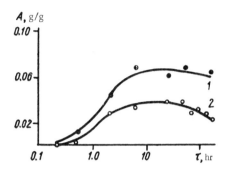

FIGURE 12. Adsorption rate of high-molecular weight polybutadiene on carbon black:

1) 665,000; 2) 930,000.

Thus no unambiguous conclusion can be drawn from the data available on the effect of the molecular weight and the concentration on the rate of adsorption of polymers from solutions.

Contradictory results were also obtained when studying the temperature dependence of adsorption. Arendt /87/ showed that the effect of temperature on the rate of establishing equilibrium is less than might be expected from diffusion laws. However, Hobden and Jellinek /74/ found that the temperature dependence of the adsorption rate can be described by

$$dG_p/dt = k\,(G_{p\infty} - G_p),\qquad (2.1)$$

where G_p is the amount of polymer adsorbed up to instant t; $G_{p\infty}$ is the equilibrium amount of the adsorbed polymer; k is the rate constant, which depends on the concentration of the solution and on the temperature. From the temperature dependence of k, the apparent activation energy of adsorption was calculated. The values obtained coincide with the activation energies for diffusion. This leads to the conclusion that adsorption kinetics are determined mainly by diffusion.

Other experimental data, especially on the adsorption of polymers on smooth surfaces, confirm this conclusion. Patat and Schliebener /88/ studied the adsorption of poly(methyl methacrylate) on aluminum and platinum surfaces. They concluded that stirring the polymer solution accelerates adsorption.

When powdered or porous materials are used, it is much more difficult to find a correlation between the adsorption rate and the diffusion process. After the outer surface has become covered, the inner surface is covered very slowly, because of the hindered diffusion into the pores /89, 90/.

The dependence of adsorption on the duration of the process is also affected by the molecular-weight distribution of the polymer, because the diffusion coefficient depends on the concentration and molecular weight. When the adsorption of the styrene-butadiene copolymer was studied, the change in the viscosity was determined together with the amount of adsorbed substance /91/. It was found that the specific viscosity first increased, and then decreased again. This shows that the adsorption proceeds with displacement of the rapidly diffusing small molecules by larger ones.

In the adsorption of poly(vinyl acetate) from benzene on wood or cellophane powders, the intrinsic viscosity – time curve passes through a maximum /92/. The same pattern was observed when the adsorption of polystyrene on carbon black was studied (Figure 13) /72/. It is interesting that the viscosity increased in the tests carried out without stirring the adsorption system. During stirring, this effect practically disappears. This is understandable, since stirring accelerates the diffusion of the macromolecules to the adsorbent.

In spite of some disagreement in the experimental data on the adsorption kinetics of polymers, it can be stated that diffusion mainly determines this kinetics. Therefore, adsorption kinetics can be found from the diffusion equation.

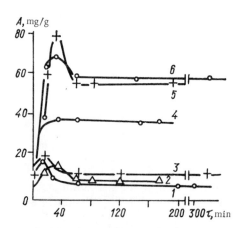

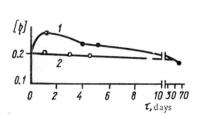

FIGURE 13. Measurement of the kinetics of the intrinsic viscosity of a polystyrene solution in CCl_4 during adsorption on carbon black:

1) tests without stirring; 2) tests with stirring.

FIGURE 14. Adsorption kinetics of solutions of lacquer PF-6 on rutile at 20°C under dynamic conditions at the following concentrations:

1) 2.7; 2) 10; 3) 5.4; 4) 40; 5) 27; 6) 15.5%.

The kinetic adsorption curves of polymers from solutions usually resemble the curves shown in Figure 6. However, kinetic curves of an unconventional type have recently been obtained /53, 93, 94/. Ermilov /93/ studied the adsorption of pentaphthalic resin (lacquer PF-6) under dynamic conditions on pigments, and found kinetic curves with a peak (Figure 14). The author believes that this shape of the kinetic curves is due to conformational changes in the molecules adsorbed on the surface.

We also obtained kinetic curves with a peak when oligo ethylene glycol adipate was adsorbed from acetone on carbon black and Aerosil silica (Figure 15). We believe that this maximum is the result of the adsorption of molecular aggregates on the solid surface followed by their destruction.

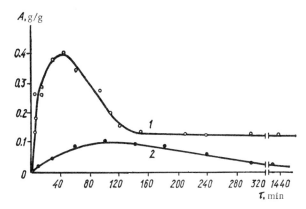

FIGURE 15. Adsorption kinetics of oligo ethylene glycol adipate from acetone on carbon black (1) and Aerosil silica (2).

Rapchinskaya and Blokh /94/ studied the adsorption of chloroprene rubber on zeolites and found that the curves rise in steps. Analogous results were obtained for fractionated polystyrene samples on carbon /74/. These were explained by the fact that the initially formed adsorbed film is reoriented after a certain period, so that the free surface is again available for further adsorption. This may also happen if the polymer is first sorbed on the outer surface of the adsorbent, and then migrates into the pores, so that it becomes possible for the polymer to be adsorbed again on the outer surface. It is worth mentioning that conventional kinetic curves were obtained for unfractionated polymers.

DESORPTION. REVERSIBILITY OF ADSORPTION

When adsorption is studied, we must take into account its reversibility, that is, the extent of desorption. It is thus possible to obtain some idea of the strength of the bond between polymer and adsorbent. Desorption is carried out by washing and drying the adsorbent after the adsorption test, or as follows. Some of the solution is withdrawn from the ampoule containing the solution of the polymer and the adsorbent plus a certain amount of adsorbed polymer, and this is replaced by pure solvent. After a certain time the ampoule is opened, and the extent of desorption is evaluated from the concentrational change in the polymer solution. When the desorption is determined under dynamic conditions, the magnitude of desorption is found by passing pure solvents through the column with the absorbent and then analyzing the eluted fractions. With some methods (contact potential method) only the irreversibly adsorbed part of the polymer can be found /95/.

Stromberg et al. /45/ studied the desorption of polyesters from the surface of glass powder and silica (Figure 16). It can be seen that the polymer is not desorbed from dilute solutions if the solvent used is the same as that employed in adsorption. However, polyesters are desorbed from silica even by the solvents from which they had been first adsorbed. This indicates that the stability of the polyester – glass bond differs from that of the polyester – silica bond. The desorption of polyesters from glass was observed only when the thermodynamic character of the desorbing solvent was more favorable than the thermodynamic character of the adsorbing solvent.

A similar pattern is found in the desorption of poly (vinyl acetate) from various surfaces. Thus, during sorption on iron powder the polymer is slowly and incompletely desorbed by carbon tetrachloride and ethylene chloride /77/, and during sorption on activated carbon, poly (vinyl acetate) is slowly and incompletely desorbed by the poor solvent acetone /89/. The use of good solvents such as acetonitrile /77/ and chloroform /89/ leads to complete desorption.

Polystyrene adsorbed on activated carbon /85/ from a poor solvent (methyl ethyl ketone) is not desorbed by this solvent, but the desorption proceeds completely in tetralin, which is a good solvent /96/. The adsorption of poly (methyl methacrylate) on carbon and alumina is irreversible in a poor solvent, but becomes reversible in a good solvent /97/.

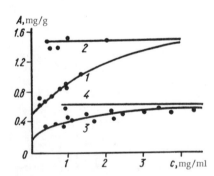

FIGURE 16. Adsorption (1, 3) and desorption (2, 4) of poly (neopentyl succinate) from toluene (1, 2) and chloroform (3, 4) on glass powder at 30°C.

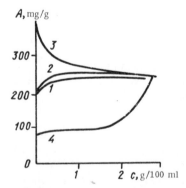

FIGURE 17. Desorption isotherms of polymeric ethylene oxide (molecular weight 4700) from carbon at 25°C:

1) adsorption isotherm from benzene; 2) desorption, benzene; 3) desorption, methanol; 4) desorption, dioxane.

Howard and McConnel /84/ studied the desorption of poly (ethylene oxide) from porous carbons after adsorption from benzene. They used three solvents for the desorption: benzene, methanol (from which the adsorption was high), and dioxane (from which the adsorption was poor). The adsorption and desorption times were the same, and therefore we cannot consider that the desorption curves were obtained under equilibrium conditions.

In desorption from benzene we observe some hysteresis (Figure 17), but after a few days the polymer is completely desorbed. When the molecular weight of the polymer is 390, the desorption proceeds rapidly, but at a higher molecular weight (190,000), greater hysteresis is observed on the sorption − desorption curves. An interesting phenomenon is noted when the polymer is desorbed by other solvents. Thus, when methanol is used instead of benzene, an amount of polymer corresponding to equilibrium adsorption from methanol passes from the solution to the adsorbent.

When dioxane is added, desorption proceeds, and 92 mg/g of polymer remain on the adsorbent, that is, an amount almost completely equivalent to the adsorption of the given polymer from dioxane solutions (93.3 mg/g).

Thus, the type of solvent considerably affects the rate and the amount of desorbed polymer. When poor solvents are used no correct conclusions on the irreversibility of adsorption can be made, and we can speak only of apparent irreversibility.

The rate and magnitude of desorption are determined not only by the quality of the solvent and its molecular weight, but also by the specific surface of the adsorbent. Figure 18 shows the sorption − desorption iso-therms of ethylene oxide copolymers of different molecular weights on nylon powder, with a small specific surface (3.4 m^2/g for nitrogen adsorption) /83/.

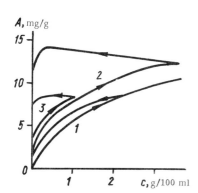

FIGURE 18. Adsorption—desorption isotherms of ethylene oxide copolymers from benzene solutions at 20°C:

1) MW 390; 2) 2980; 3) 190,000.

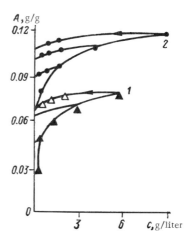

FIGURE 19. Hysteresis on the adsorption—desorption curves in the styrene-butadiene copolymer—carbon black system:

1) carbon black with specific surface 45.6; 2) 75.1 m^2/g.

At all molecular weights, the hysteresis on the curves increases with increase in molecular weight.

The polymer is desorbed more rapidly when other solvents are used. Thus, desorption in dioxane is 30%, and in methanol or dimethylformamide it is 90%. For the highest molecular weight (190,000), desorption is complete when excess methanol is added, apparently because of preferential adsorption of methanol on the nylon surface.

In the adsorption of polyisobutylene from hexane, the content of irreversibly sorbed polymer increases with increase in the molecular weight of the polymer: polyisobutylene with a molecular weight of 8800 is 39% desorbed, while with a molecular weight of 32,500 it is only 3% desorbed /84/.

On solvents with a large specific surface, the desorption rate is low, and frequently adsorption is irreversible, although different solvents with various properties are used. Kraus and Dugone /76/ studied the adsorption and desorption of the styrene-butadiene copolymer on carbon black, and observed large hysteresis loops on the adsorption − desorption curves (Figure 19). To find the irreversibility limit after the adsorption tests, the polymer was extracted from heptane solutions for 72 hours in a Soxhlet extractor. When heptane was used as the desorbent, the amount of unextracted polymer was approximately equal to the amount obtained in Figure 19 by extrapolating the desorption curve to zero concentration. When benzene was used, a larger amount of the polymer was desorbed, but not the whole amount.

Evidently, the adsorption process takes place in two stages: rapid physical adsorption of the polymer, followed by a slow reaction of the polymer with carbon black, that is, a chemical bond is formed between the molecules and the surface. Naturally, in the case of chemisorption, desorption will be incomplete.

Irreversible adsorption is most frequently observed on porous sorbents. For example, the desorption of polystyrene from activated carbon proceeds very slowly, and is mainly irreversible /98/. The desorption of polyesters from carbon is also slight /97/.

However, adsorption on smooth surfaces and on adsorbents with a small specific surface is reversible. Thus, the adsorption of polyesters with molecular weights between 970 and 6250 is reversible on glass powder /49/. Poly(ethylene glycol) with molecular weight 28,000 is reversibly adsorbed on aluminum and glass film from benzene and methanol /99/. Desorption by pure solvents is complete, and adsorption is dependent on temperature and concentrations.

Patat and Schliebener /88/ studied the adsorption and desorption of polymers on smooth surfaces from concentrated solutions (up to 4%). In most cases the authors observed a high degree of desorption. In some cases the time of desorption was shorter than the time of adsorption, which indicates that the bond between polymer and adsorbent is weak. For example, the desorption of polystyrene in benzene is 100%, in butanone 97%, and in cyclohexanone 90−95%. It was found that for poly(methyl methacrylate) and other polymers, desorption increases with increase in the concentration of the solution from which the polymer was adsorbed, the amount of adsorbed substance, and the temperature.

Tests on the adsorption and desorption rates thus indicate that these rates differ on smooth and porous adsorbents.

Moreover, if the polymer was adsorbed from dilute solutions, desorption is slow and not always complete, while after adsorption from concentrated solutions, desorption is rapid and frequently complete. This is apparently due to differences in the stability of the adsorption bonds arising during adsorption from dilute and concentrated solutions. In the latter case the adsorption bonds are weak, as indicated by the rate and magnitude of adsorption. Hence, during adsorption, the polymer can be bound reversibly or irreversibly, while the degree of irreversibility depends on the nature of the solvent, molecular weight of adsorbent, and type of adsorbent surface. No detailed studies on the reasons for such irreversibility can be found in the literature.

APPLICATION OF KINETIC EQUATIONS FOR DESCRIBING ADSORPTION AND DESORPTION PROCESSES

Real kinetic functions for sorption and desorption processes are not always correctly described by simple kinetic equations, since they do not allow for the complexity of the processes occurring during polymer adsorption. However, by applying conventional kinetic equations to adsorption and desorption processes it is possible to describe these processes rather approximately.

Peterson and Kwei /81/ assume that the adsorption of poly (vinyl acetate) from very dilute solutions on smooth surfaces can be described by

$$\frac{d\Theta}{d\tau} = K_1 (1 - \Theta) c - K_2 \Theta, \tag{2.2}$$

where Θ is the fraction of total surface coverage; c is the concentration of the polymer solution; and K_1 and K_2 are the rate constants of the adsorption and desorption processes. If the adsorbed molecules do not interact, Θ is directly proportional to the amount of adsorbed molecules (n):

$$\Theta = n/N, \tag{2.3}$$

where N is the value of n for total single-layer surface coverage; the value of N is $1 \cdot 10^{-9}$ mole/cm^2 if we assume that the plateau on curves 1–3 (see Figure 8) corresponds to saturation adsorption by the monolayer. When $d\Theta/dt = 0$, then

$$\Theta_S = \frac{n_{plateau}}{N}. \tag{2.4}$$

Θ_S calculated for different concentrations varied between 0.8 and 0.96. From equation (2.2) it follows that at $d\Theta/dt = 0$,

$$\frac{c}{\Theta_S} = c + \frac{K_2}{K_1}. \tag{2.5}$$

The curve of $c/\Theta_S = f(c)$ is a straight line with a slope of 45°, and the intercept on the ordinates is equal to K_2/K_1. By integrating (2.2) we obtain

$$-\ln\left[1 - \left(1 + \frac{K_2}{K_1 c}\right)\Theta\right] = K_1 c\tau. \tag{2.6}$$

The dependence of $\log[1-(1 + K_2/K_1 c)\Theta]$ on time τ is described by a straight line with a slope of $K_1 c/2.303$ (Figure 20). Figure 20 shows that the experimental points lie on a straight line, which confirms that (2.2) can be applied to describe the adsorption of a polymer (in particular, poly(vinyl acetate)) from dilute solutions on smooth surfaces.

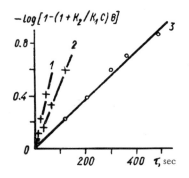

FIGURE 20. Kinetic curves of the adsorption of poly(vinyl acetate) described by equation (2.5):

1) concentration of solution $1.15 \cdot 10^{-4}$; 2) $5.75 \cdot 10^{-5}$; 3) $2.30 \cdot 10^{-5}$ mole/liter.

Assuming that the adsorption isotherm is described by Langmuir's equation, Peterson and Kwei /81/ postulated that the concentration of the polymer solution remains constant during the adsorption process. However, this condition holds in few cases only. Therefore, Jankovics /100/ attempted to describe the adsorption kinetics under conditions where the initial concentration of the polymer solution changes appreciably during the absorption process. In this case $(c_0 - \alpha\Theta)$ must be substituted for c in (2.2), where c_0 is the initial concentration of the polymer and α is a constant:

$$d\Theta/d\tau = K_1(c_0 - \alpha\Theta)(1 - \Theta) - K_2\Theta, \tag{2.7}$$

or

$$d\Theta/d\tau = K_1[\alpha\Theta^2 - \Theta(c_0 + \alpha + b) + c_0], \tag{2.8}$$
$$b = K_2/K_1.$$

After integration, the following equation is obtained:

$$K_1\tau = \frac{1}{(c_0 - \alpha + b)} \ln\{[(c_0 + b) - \alpha\Theta]/[(c_0 + b)(1 + \Theta)]\}. \qquad (2.9)$$

It is believed that (2.9) can be applied when the adsorption isotherm can be described by Langmuir's equation. This equation was verified on the system calcium phosphate—polyacrylamide, since Langmuir isotherms were obtained for it /101/.

Aqueous solutions of polyacrylamide with different molecular weights were used (Figure 21). When the applicability of (2.9) to the given system was checked, the authors assumed that the time dependence of surface coverage is $\Theta A/A_p$, where A is the amount adsorbed at instant τ, and A_p is the amount adsorbed under equilibrium conditions. Then the dependence of $\log[(c_0 + b)A_p - \alpha_A/[(c_0 + b)(A_p - A)]$ on τ was plotted, and should be a straight line.

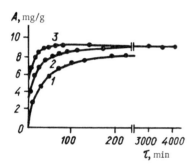

FIGURE 21. Adsorption kinetics of poly-acrylamide:

1) MW $6.0 \cdot 10^6$; 2) $3.0 \cdot 10^6$; 3) $1.0 \cdot 10^6$ (initial concentration 0.1%).

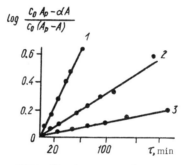

FIGURE 22. Kinetic curves for the polyacrylamide—calcium phosphate system given by equation (2.9):

1) MW $1.0 \cdot 10^6$; 2) $3.0 \cdot 10^6$; 3) $6.0 \cdot 10^6$.

It is assumed that the desorption rate is much lower than the adsorption rate ($K_2 \ll K_1$), and therefore b can be neglected. This assumption is correct if the polymer − adsorbent interaction is strong, as observed in the system studied. Figure 22 shows the dependence of $[c_0A_p - \alpha A]/[c_0(A_p - A)]$ on τ at $\alpha = A_p$. From the slopes of the straight lines the rate constants were obtained. They are between $5.46 \cdot 10^3$ and $3.09 \cdot 10^3$ $sec^{-1} \cdot mole^{-1}$, and decrease with increase in the molecular weight of the polymer. Peterson and Kwei /81/ showed that equation (2.9) can also be used for other systems, even when the adsorption isotherms are not described by Langmuir's equation.

Howard and McConnel /84/ attempted to apply the kinetic equations of Kipling /102/ to the adsorption of poly(ethylene oxide) with molecular weights 190,000 and 4700 on porous adsorbents (carbon and nylon) from benzene solutions. The equation $A_t = A_\infty [1-\exp(-k\tau)]$ did not hold for the experimental data. The systems studied are described by equation $A_t = A_\infty \tau/(\tau + k)$, where k is the time of contact during which 50% of the amount of polymer studied is adsorbed. It was found that the adsorption rate is lower on nylon ($k = 3.5$ hr) than on carbon ($k = 0.12$ hr), because of the lower diffusion rate of the macromolecules into the narrow pores of nylon.

Polonskii and Zakordonskii /121/ studied the adsorption kinetics in the system β-cyanoethyl esters of poly(vinyl alcohol) — powderlike adsorbents. They found that the experimental data corresponding to the initial section of the kinetic curve can be described by an equation of type

$$v = k\tau^{-n}. \qquad (2.10)$$

Exponent n given by the conditions of diffusion of the macromolecules to the adsorbent surface was found to be constant for all the systems, but coefficient K depended on the nature of the polymer. Besides the diffusion rate, we must take into account the rate of establishing the equilibrium conformation of the adsorbed macromolecule which is given by

$$v_2 = k_2 f (1 - \Theta), \qquad (2.11)$$

where k_2 is the equilibrium constant; f is a function characterizing the flexibility of the macrochain; Θ is the fraction of the surface covered by the polymer molecules.

The kinetic features of the process of adsorption of macromolecules is thus determined not only by the equations of diffusion of these molecules to the adsorbent surface, but also depend on the character of the secondary processes of the redistribution of macromolecules and individual polymeric segments in the adsorption system.

Chapter 3

THE FUNDAMENTAL LAWS OF THE
ADSORPTION OF POLYMERS

EFFECT OF THE THERMODYNAMIC CHARACTER
OF THE SOLVENT (SOLVENT POWER)
ON ADSORPTION

The nature of the solvent is one of the most important factors determining polymer adsorption. It is clear that the conformation of the polymeric chain in the solution depends appreciably on the nature of the solvent. The solvent determines the size of the macromolecule in the solution and the asymmetry of the polymeric coil. These factors determine the conditions of contact of the polymeric molecule with the surface, the possible orientation of the macromolecule on the surface, the structure of the adsorption layer, and other parameters. The adsorption surface of the solvent itself is of great importance, as it may sometimes be the principal factor, and thus lead to incorrect ideas on the probable character of the reaction of the polymeric molecule with the adsorbent surface.

The polymeric molecule is in the form of a coiled conformation in the dilute solution. The size and shape of this coil depend on the energy of interaction between the polymeric and solvent molecules, and on intramolecular interactions.

There is a well-defined correlation between the polymer — solvent interaction, defined in the Flory-Huggins theory by interaction parameter χ, and the chain size. The correlation is based on the concept of the osmotic action of the solvent on the polymeric molecule present in the shape of a statistical coil /103/. Because of the osmotic action of the solvent, the coil swells and becomes inflated, and the molecule passes into a state of less probable conformation. This state is determined by the equilibrium between the osmotic forces attempting to expand the molecule and the elastic forces that hinder such expansion. The osmotic pressure of polymer solutions is given by

$$\frac{\pi}{c} = RT/M + A_2 c + A_3 c^2 + \cdots, \qquad (3.1)$$

where π is the osmotic pressure; c is the concentration of the solution; M is the molecular weight of the polymer; A_2 and A_3 are virial coefficients. The second virial coefficient A_2 characterizes the energy of interaction between the molecules of the solvent and, with parameter χ, can serve as the criterion of the solvent power. At higher values of A_2, the solvent is better. A decrease in the second virial coefficient leads to a decrease in the osmotic pressure and hence a decrease in the size of the macro-molecules in the solution. This in turn leads to a lower viscosity of the solution. Thus, the viscosity is related to the solvent power.

Schulz /104, 105/ compared the different methods for evaluating the quality of the solution, and showed that for many polymers the second virial coefficient varies symbatically with the intrinsic viscosity $[\eta]$. Therefore some authors use $[\eta]$ as a criterion of the power of the solvent.

We shall discuss some experimental data on the adsorption of polymers in solvents of different powers.

In dilute solutions of polymers in poor solvents, the macromolecules are coiled more if their size is smaller, and are usually adsorbed more than from good solvents because of the weaker interaction with the solvent. Figure 23 shows the adsorption isotherms of polyesters from good (chloroform), and poor (toluene) solvents. The adsorption from a poor solvent is 2 to 4 times that from a good one.

Koral et al. /77/ studied the adsorption of poly(vinyl acetate) on iron from various solvents and showed that in most cases the amount of the adsorbed polymer is inversely proportional to the intrinsic viscosity, that is, the adsorption is greater from a poor solvent (Figure 24).   On carbon, poly(vinyl acetate) is adsorbed most from methanol, and then in descending order from toluene, benzene, acetone, trichloroethylene, and 1,2-dichloroethane. From chloroform, in which poly(vinyl acetate) has the highest intrinsic viscosity, no adsorption is observed /89/.

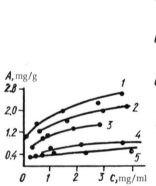

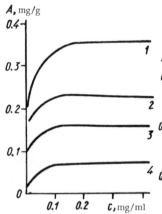

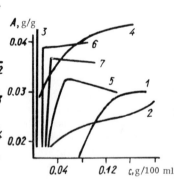

FIGURE 23. Adsorption iso-therms of polyesters from toluene (1, 2) and chloroform (4, 5) on glass at 30°C:

1) poly(trimethylene adipate); 2,4) poly(pentamethylene succi-nate); 3) poly(propylene adipate); 5) poly(ethylene adipate).

FIGURE 24. Adsorption iso-therms of poly(vinyl acetate) on glass from different solvents at 30°C:

1) carbon tetrachloride; 2) toluene; 3) benzene; 4) ethylene chloride.

FIGURE 25. Adsorption isotherms of GRS rubber on carbon black:

1) benzene; 2) chloroform; 3) n-heptane; 4) chloroform + n-heptane; 5) ethanol + chloroform (9:1); 6) chloroform + ethanol (8:2); 7) benzene + ethanol (9:1).

TABLE 1. Intrinsic viscosity and maximum adsorption of poly(vinyl acetate)

Solvent	[η]	A, mg/g
Chloroform	1.133	0.070
Benzene	0.795	0.158
Toluene	0.554	0.227
Carbon tetrachloride	0.281	0.335

Figure 24 and Table 1 show that the highest adsorption is observed in the solvent of the lowest power, namely, carbon tetrachloride. The adsorption of the styrene-butadiene copolymer on carbon black /106/ is also greater from a solvent which has a lower intrinsic viscosity, namely n-heptane. The addition of a precipitating agent (ethanol) to benzene or chloroform, which lowers the power of the solvent and decreases the intrinsic viscosity, improves the adsorption (Figure 25).

The copolymer of isobutylene and isoprene is adsorbed on carbon black much better from benzene solutions (good solvent) than from cyclohexanone solutions (poor solvent) /107/.

The adsorption of acetylcellulose on starch from acetone solutions is better than it is from dioxane solutions /108/. For nitrocellulose, the amount adsorbed decreases in the sequence acetone < ethanol < dioxane. The addition of a poor solvent to a good one increases adsorption. For example, the adsorption of nitrocellulose from a cyclohexane + acetone mixture increases with increase in the acetone level, from cyclohexane + methanol with increase in the cyclohexane level, and from acetone + methanol with increase in acetone level /109/. Solvent power decreases in the order acetone, cyclohexane, ethanol, dioxane. Hence, larger amounts were adsorbed from the poorer solvent in all cases. An exception is the acetone − water system, for which the amount of adsorbed polymer increases with increase in the water level, although water is a precipitating agent for cellulose.

Polystyrene on activated carbon, silica, and alumina is better sorbed from the poorer solvent, toluene, than from the better solvent, dichloro-ethane /110/. Figure 26 shows the adsorption isotherms of polystyrene from different solvents. The parameter χ for the interaction between polymer and solvent for the system polystyrene − butanone is 0.485, and for polystyrene in toluene it is 0.434, that is, the solvent power of butanone is lower. Hence, adsorption from butanone is greater than from toluene.

The solubility parameter δ is also used as a criterion for the solvent power. When the difference in δ for polymer and solvent is greater, the solvent is poorer. Mizukara et al. /47/ compared the values of the adsorption and the solubility parameter (Figure 27), and found that there is a linear relationship between these magnitudes.

Hara and Imoto /111/ studied the adsorption of the copolymer of vinyl acetate and ethylene on glass from different solvents (Figure 28). The authors tried to explain the dependence of the adsorption on the solvent power by the change in the area occupied by the polymeric chain on the adsorbent surface.

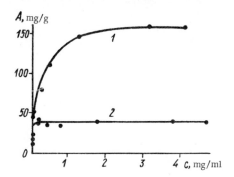

FIGURE 26. Adsorption isotherms of polystyrene from butanone (1) and toluene (2) on carbon.

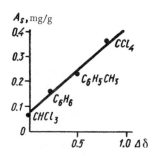

FIGURE 27. Dependence of the maximum adsorption on the difference between the solubility parameters of poly(vinyl acetate) and the solvent.

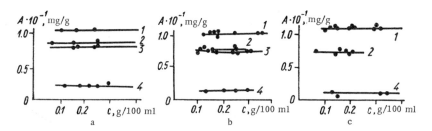

FIGURE 28. Adsorption isotherms of the ethylene−vinyl acetate copolymer of molecular weight (M) $12.0 \cdot 10^4$ containing 29% vinyl acetate (a); $M \leqslant 12.0 \cdot 10^4, 30.2\%$ vinyl acetate (b); $M = 12.0 \cdot 10^4, 14.7\%$ vinyl acetate (c) on glass from different solvents:

1) carbon disulfide; 2) benzene; 3) carbon tetrachloride; 4) trichloroethylene.

We know that the radius of gyration of the polymer chain is larger in a good solvent than in a poor one. The size of the isolated chain can be estimated from Flory's equation

$$|\eta|\, M = \Phi' \, \langle \bar{S}^2 \rangle^{\frac{3}{2}}, \qquad (3.2)$$

where $[\eta]$ is the intrinsic viscosity; M is the molecular weight; Φ' is Flory's constant; $\langle \bar{S}^2 \rangle^{\frac{1}{2}}$ is the radius of gyration.

The cross-sectional area (a) of the polymeric coil in the solution at constant temperature is

$$a \simeq (\langle \bar{S}^2 \rangle^{\frac{1}{2}})^2 \simeq [\eta]^{\frac{2}{3}}. \qquad (3.3)$$

If we postulate that the maximum adsorption (A) is directly dependent on the elongation of the polymeric chain, then the relationship between the maximum adsorption and the intrinsic viscosity will be

$$A \simeq k'/[\eta]^{-\frac{2}{3}} f([\eta]).$$ (3.4)

For the system poly(vinyl acetate)—glass, the relationship between the maximum adsorption and the degree of chain elongation can be written (Figure 29) as

$$A_S = k^1/[\eta]^{-\frac{2}{3}[\eta]}.$$ (3.5)

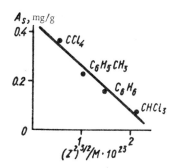

FIGURE 29. Dependence of maximum adsorption on the specific volume of the poly(vinyl acetate) molecule in solution.

Table 2 shows the values of A_S and $[\eta]$ for the system ethylene-vinyl acetate copolymer — glass.

TABLE 2. Maximum values of the adsorption and intrinsic viscosity at 30°C for the system ethylene-vinyl acetate polymer (MW = 120,000)—glass

Vinyl acetate concen- tration, %	CS_2		C_6H_6		$CHCl_3$		CCl_4	
	$[\eta]$	A_s, mg/g	$[\eta]$	A_s, mg/g	$[\eta]$	A_s, mg/g	$[\eta]$	A_s, mg/g
29.0	0.925	0.106	1.148	0.085	1.369	0.020	1.311	0.080
30.2	0.522	0.105	0.742	0.079	0.805	0.014	0.752	0.077
14.7	0.752	0.109	0.847	0.084	0.918	0.11	—	—

Equation (3.5) is applicable to the adsorption of poly(vinyl acetate) on glass, when we can assume a plane chain configuration /111/. This, however, is not true in the adsorption of the ethylene-vinyl acetate copolymer on glass. In this case the polymeric chain is sorbed as a loop on the adsorbent surface (Figures 29 and 30a). If we assume that the area occupied by the polymeric chain adsorbed as a loop is proportional to the area occupied by the molecule in the solution, then equation (3.5) can be written in the form

$$As = \tau/[\eta]^{\frac{2}{3}},\qquad\qquad (3.6)$$

where τ is constant. From experimental data on the adsorption of the ethylene-vinyl acetate copolymer, presented in the form (3.6) (Figure 30b), we can see that this equation holds except in adsorption from chloroform.

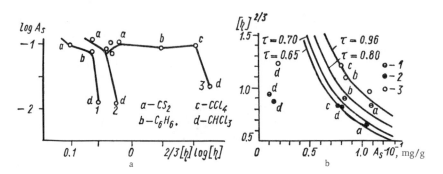

FIGURE 30. Relationship between A_S and the area occupied by the adsorbed molecule:

1) 14.7; 2) 30.2; 3) 29.0% vinyl acetate in the copolymer.

Hence, when the polymer is adsorbed from dilute solutions, a distinct correlation between the maximum adsorption and intrinsic viscosity was established in the systems studied. The intrinsic viscosity itself depends on the size of the polymeric chain in the solution.

In the adsorption of polymers from different surfaces we mainly observe strong adsorption from solvents of low solvent power. This is analogous to the adsorption of low-molecular weight substances. Sometimes, however, the pattern is reversed, and the polymers are more strongly adsorbed from good solvents. Thus, poly(dimethyl siloxane) on carbon and pigments is adsorbed more from carbon tetrachloride than from xylene /50/. However, the intrinsic viscosity of the polymer is higher in carbon tetra-chloride than in xylene, while the solubility parameter of poly(dimethyl siloxane) is closer to that of CCl_4 than to that of xylene.

The pattern of the adsorption of poly(methyl methacrylate) on glass and iron powder from different solvents is more complex /112/. Table 3 shows that it is impossible to find any correlation between adsorption and the parameters characterizing the solvent. A study of the adsorption of poly(methyl methacrylate) from a solvent mixture showed that sometimes the addition of a precipitant to the solvent leads to reduced adsorption, in spite of deterioration of the solvent. This can be clearly seen in Figure 31, which shows the dependence of the adsorption on the mole fraction of the precipitatnt in the mixture.

The adsorption of poly(methyl trisiloxane) on glass from a mixture of acetonitrile (precipitant) and benzene (good solvent) is also much less than it is from the pure solution (Figure 32). When powder is used, the polymer is not adsorbed at all from the mixture /54/.

TABLE 3. Maximum adsorption of poly(methyl methacrylate) on glass in different solvents

Solvent	Dielectric constant	Dipole moment	Polarizability	Solubility parameter	A_s, mg/g
Toluene	2.44	0.36	31.5	8.90	0.86
trans-Dichloroethane	2.14	5	—	9.0	0.72
Benzene	2.28	0	27.1	9.15	0.65
Trichloroethylene	3.4	—	—	5.3	0.65
cis-Dichloroethane	9.20	1.90	19.9	9.1	0.58
1,2-Dichloroethane	2.14	1.24	22.0	9.8	0.45
Methylene chloride	9.08	1.57	20.0	9.7	0.42
1,1,2-Trichloroethane	—	1.22	33.6	9.6	0.35
1,1,2,2-Tetrachloroethylene. . . .	8.2	1.36	32.4	9.3	0.34
Chloroform	4.8	1.02	24.8	9.3	0.21

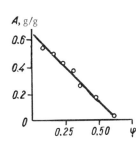

FIGURE 31. Dependence of the adsorption of poly(methyl methacrylate) from a benzene+ acetonitrile mixture on the mole fraction of acetonitrile in the solution.

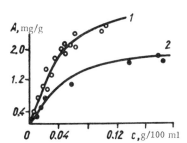
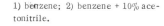

FIGURE 32. Adsorption isotherms of poly(methyl siloxane) (MW = 536,000) on glass from different solvents:

1) benzene; 2) benzene + 10% acetonitrile.

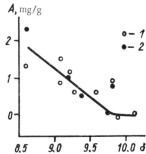

FIGURE 33. Dependence of the adsorption of poly (vinyl acetate) on the solubility parameter of the solvent:

1) cellulose; 2) iron powder.

Luce and Robertson /113/ found that the adsorption of poly(vinyl acetate) on cellulose depends on the nature of the solvent. The maximum values of the adsorption and the intrinsic viscosities (Table 4) show that there is no distinct relationship between these parameters. However, the authors of /113/ attempted to find a relationship between the solubility parameter and the adsorption from different solutions (Figure 33). A certain correlation could be established, but no clear sequence could be found for all the solvents.

The above results show that when the effect of the solvent on the adsorption of a polymer is considered, the interaction of both the polymer and the solvent with the surface of the adsorbent is as important as the polymer − solvent interaction. A strong solvent − surface interaction will decrease the adsorption of the polymer, and may prevent transition of the macromolecules onto the adsorbent. This may explain the lower adsorption of poly(dimethyl siloxane) from an acetonitrile − benzene mixture /54/ since acetonitrile has a strong affinity to glass.

TABLE 4. Adsorption of poly(vinyl acetate) on cellulose from different solvents

Solvent	$[\eta]$	A, mg/g
Carbon tetrachloride	32	59.4
Benzene	90	45.2
Ethylene chloride	97	38.5
Methyl ethyl ketone	81	28.0
Ethyl acetate	86	37.6
Nitrobenzene	—	25.3
Dioxane	84	-0.8
Acetone	82	-3.0

Howard and McConnel /84/ studied the adsorption of ethylene oxide copolymers on solid surfaces from different solvents, and the affinity of these solvents to various adsorbents. Ethylene oxide polymers were adsorbed on Aerosil silica from chloroform, methanol, water, dioxane, and dimethylformamide (Figure 34). No adsorption was observed from the last two solvents. The adsorption changes in the following sequence: chloroform > water > methanol > dioxane > dimethylformamide. Since the solvent power can be characterized by the intrinsic viscosity, Howard and McConnel determined $[\eta]$ for the given solutions, and found the following sequence: chloroform > benzene ~ dioxane ~ dimethylformamide ~ water > > methanol.

The authors also determined the parameters of the polymer — solvent interaction. These change in the following order: chloroform > benzene > dioxane > methanol. If the magnitudes characterizing the power of the solvents and the adsorption of polymers from them are compared, no distinct correlation can be observed between them. This is because when the dependence of adsorption on the nature of the solvent is considered, we must allow for the interaction of the solvents with the adsorbent.

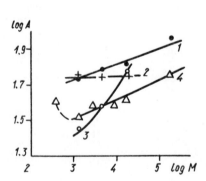

FIGURE 34. Adsorption isotherms of the ethylene oxide copolymer on Aerosil silica:

1) from chloroform; 2) from water; 3) from methanol; 4) from benzene.

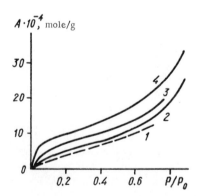

FIGURE 35. Adsorption isotherms of the vapors of chloroform (1), benzene (2), dioxane (3), and methanol (4) on Aerosil silica.

We obtained the sorption isotherms of the vapors of the given solvents on Aerosil silica (Figure 35), and from them determined parameter **c** in the BET equation, and the heats of adsorption (Table 5) which qualitatively (but not quantitatively) agree with the published data /114/. The sorption isotherms of dimethylformamide (DMF) by Aerosil silica were not obtained, but tests showed that the affinity between DMF and Aerosil silica is high. It was found that the addition of DMF to a solution of the given polymer markedly reduces the adsorption, probably because of the strong adsorption of polar DMF on Aerosil silica. There is no doubt that water will also strongly reduce adsorption on Aerosil.

TABLE 5. Parameters of the BET equation for the adsorption of solvent vapors by Aerosil silica at 20°C

Vapor	c	$E_A - E_L$, cal/mole
Methanol	99.0	2.7
Dioxane	65.0	2.4
Benzene	9.5	1.4
Chloroform	5.1	1.0

Depending on the affinity of solvents to Aerosil silica, the following sequence can be written: dimethylformamide > methanol > dioxane > > benzene > chloroform. If we know the power of the solvents and their affinity to the adsorbent, the adsorption data can be explained as follows. The ethylene oxide copolymers are adsorbed slightly on Aerosil silica from DMF solution because of the strong affinity of the solvent to the adsorbent. The adsorption of these polymers from methanol is also slight for the same reason, but in this case the adsorption is strongly dependent on the molecular weight. Water has a medium solvent power for ethylene oxide copolymers and is certainly appreciably adsorbed on the Aerosil silica surface. Nevertheless, adsorption from water is stronger than from methanol. This may be due to the fact that the polar Aerosil silica surface is partially covered by water molecules and can adsorb the macromolecules, while the methanol molecules block the active centers on the adsorbent surface and prevent adsorption. The high solubility of the given polymers in chloroform does not inhibit their sorption from the solvent, since the affinity of chloroform to Aerosil silica is slight. The solvent power of benzene is poor, but this solvent is strongly adsorbed on Aerosil silica.

A comparison of the adsorption from these two solvents shows that in the given case the adsorbent − solvent interaction is the decisive factor. We did not study the reason why polymers are not adsorbed from dioxane. After the adsorption of polymers of ethylene oxide from dioxane on a porous adsorbent (charcoal) had been investigated, we established the following sequence of adsorption: water > methanol > benzene > dioxane > >chloroform > DMF. It can be seen that this sequence differs from that in the adsorption of the same polymer on Aerosil silica, but a relationship is noted between the adsorption and the power of the solvent for high-molecular weight polymers. The adsorption from the better solvent (chloroform) is less than from the poorer one (methanol).

The affinity of the solvent to the adsorbent can be arranged in the following sequence: DMF > dioxane > chloroform > benzene > methanol > > water. These data were obtained by sorption tests of the solvent vapor on charcoal.

Although the solvent power of water is similar to that of DMF, the adsorption from them differs, because their affinity to the adsorbent surface differs. Methanol is a poor solvent for high-molecular weight ethylene oxide polymers, while its affinity to the carbon surface is low. For this reason the adsorption from methanol is considerable. The adsorption from chloroform is slight, since this solvent has a high affinity to both the adsorbent and to the polymers studied. The polymer is appreciably adsorbed from benzene because although this solvent is poor, it has a low affinity to the adsorbent surface. However, the adsorption from dioxane is much weaker than from benzene, because although their solvent powers are similar, their affinities to the carbon surface differ.

On graphitized channel black, which is a nonporous and nonpolar adsorbent, ethylene oxide polymers are sorbed only from solvents with a slight affinity to the adsorbent surface, namely, from methanol and water /84/. The affinity of the solvent to the adsorbent is also important in the adsorption of these polymers on nylon. The polymer can be adsorbed from benzene, which has the lowest affinity to nylon.

By comparing the parameters characterizing the power of the solvent and its affinity to the adsorbent surface it is possible to give some explanation of the dependence of the polymer adsorption on the nature of the solvent. However, even in this case no clear relationships can always be found. This can be seen from the data of Howard and McConnel /84/ (adsorption from dioxane and benzene, with similar solvent powers and affinities to the adsorbent surface). It is probable that structure formation phenomena in polymer solutions, which depend on the nature of the solvent, that is, on the type of polymer – polymer interaction, are important in adsorption. These processes become apparent in more concentrated solutions.

EFFECT OF TEMPERATURE ON ADSORPTION

Temperature variations during adsorption increase the mobility of the macromolecules, change the solvent power, and also affect the adsorption of the solvent competing with the adsorption of the polymer. Hence, an increase in temperature has an appreciable effect on adsorption.

Stromberg /45/ studied the adsorption of polyesters on glass at different temperatures. He found that adsorption decreases in the poly-(neopentyl silicate) – toluene – glass system when the temperature increases from 30 to 87°C. The author connects this phenomenon with improvement in solution. At a higher solvent power, the adsorption decreased for the given polymer. In another solvent, chloroform, a temperature increase from from 30 to 50°C did not change the adsorption. Evidently the solvent power was practically temperature independent. Perkel and Ullman /54/, like Stromberg, correlate the temperature dependence of the adsorption with

solvent power, but they do not report any experimental data to support this
hypothesis. The adsorption of the styrene-butadiene copolymer also
decreases with temperature; this phenomenon is more marked for the
copolymer of higher molecular weight /107/.

The adsorption of one and the same polymer may have a different
temperature coefficient, depending on the nature of the solvent or adsorbent.
Table 6 shows that the adsorption of poly(methyl methacrylate) on iron
powder decreases with increase in temperature /112/. However, the
adsorption of poly(methyl methacrylate) on glass from benzene does not
change when the temperature is increased from 30 to 70°C.

TABLE 6.   Adsorption of poly(methyl methacrylate) on iron powder

MW of polymer $\times 10^{-3}$	Solvent	t, °C	A_S, mg/g	MW of polymer $\times 10^{-3}$	Solvent	t, °C	A_S, mg/g
23	Benzene	8	0.55	2270	Chloroform	68	0.21
23	"	52	0.50	2270	Toluene	30	0.84
118	"	8	0.60	2270	"	68	0.72
118	"	52	0.54	2270	"	105	0.64
532	"	30	0.75	532*	Benzene	30—70	0.70
532	"	70	0.70	532*	"	70—30	0.69
2270	"	30	0.65	2270*	"	30—110	0.77
2270	"	52	0.60	2270*	"	110—30	0.79
2270	"	68	0.56	2270*	Chloroform	30—68	0.22
2270	Chloroform	30	0.27	2270*	Toluene	68—30	0.73

* The polymer solution with the adsorbent was shaken at temperature t_1 and then at temperature t_2.

Ellerstein and Ullman /112/ carried out several tests to study the
temperature dependence of adsorption under the following conditions. The
solution of the polymer and the adsorbent was held at temperature t_1 until
equilibrium was established (continuous mixing); the temperature in the
thermostat was then raised to t_2, and the adsorption was determined. The
results were compared with those in the conventional determination of
adsorption at t_2. It was found that if $t_2 > t_1$, the amount of the polymer
adsorbed is independent of the experimental conditions. However, if
$t_2 < t_1$, the adsorption at both temperatures is equal (see Table 6). It
appears that adsorption equilibrium was not established in the latter case.

The adsorption of poly(methyl methacrylate) from toluene or dioxane
solutions on glass, aluminum, and silica increases with temperature,
the adsorption of polystyrene on these adsorbents is practically temperature
independent, while the adsorption of poly(vinyl chloride) decreases with
increase in temperature /110/. When the adsorption of these polymers
was studied on carbon, the following changes were observed: the
adsorption of poly(methyl methacrylate) decreases with temperature, that
of poly(vinyl chloride) increases, while the adsorption of polystyrene is
practically temperature independent.

In some cases a considerable increase in polymer adsorption with temperature was observed, for example, in the adsorption of poly(methacrylic acid) on alumina /115/, or in the adsorption of SKN-26 rubber on ferric oxide from toluene solutions (Figure 36) /116/. Soltys et al. /115/ explain the increase in the adsorption of poly(methacrylic acid) by the deterioration in the solvent power with increase in temperature. This is confirmed by the decrease in the intrinsic viscosity. At the same time, a temperature rise changes the flexibility of the polymeric chain, which also affects the adsorption power. The increase in the adsorption of the SKN-18 and SKN-26 rubbers might be due to thermal deaggregation of the molecules, so that the rubber macromolecules can penetrate more readily into the pores of the sorbent /116/.

Figure 37 shows the dependence of the equilibrium concentration on the reciprocal of the absolute temperature /113/ in the poly(vinyl acetate)—benzene — cellulose system. In this case a considerable increase in the adsorption with temperature is observed.

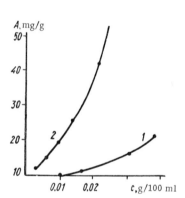

FIGURE 36. Adsorption isotherms of SKN-26 rubber from toluene solution on ferric oxide at 12°C (1) and 22°C (2).

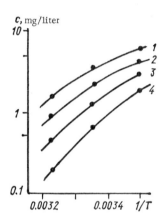

FIGURE 37. Temperature dependence of the adsorption of poly(vinyl acetate) from benzene on cellulose;

1) $A = 40$; 2) 35; 3) 30; 4) 25 mg/g.

A slight increase in adsorption with temperature is observed in the poly(isoproyyl acrylate)— chloroform — silica system /117/. The adsorption of ED-6 epoxide resin on powdered silica gel also increases with temperature /118/.

Patat et al. /88/ studied the temperature dependence of the adsorption of some copolymers from concentrated solutions (up to 4%) on smooth surfaces. They found a minimum of adsorption and also a monotonic increase or decrease of the adsorption with temperature. Table 7 shows the temperature dependence of the adsorption for poly(vinyl acetate) on copper and platinum foil and for polyvinylpyrrolidone on platinum.

TABLE 7. Temperature dependence of the adsorption of polymers

Adsorbent	Solvent	t, °C	A_S, mg/g
		Poly (vinyl acetate)	
Copper foil	Benzene	12.2	21.0
The same	"	17.6	16.6
"	"	25.1	8.7
"	"	35.2	18.5
"	"	45.4	28.0
Platinum	Toluene	12.2	18.3
"	"	18.6	8.9
"	"	25.1	20.8
"	"	35.3	18.6
		Polyvinylpyrrolidone	
"	Toluene	15.5	184.6
"	"	20.2	119.6
"	"	25.2	106.9
"	"	30.3	106.8
"	"	30.1	114.0
"	"	35.1	145.8

Interesting results were obtained when the temperature dependence of the adsorption was studied for poly (ethylene glycols) of different molecular weights from benzene solutions on aluminum (Figure 38) /99/. The adsorption is almost temperature independent at low concentrations of the solutions.

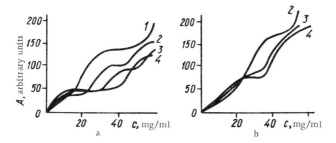

FIGURE 38. Adsorption isotherms of poly(ethylene glycol) of MW = 6000 (a) and 28,000 (b) from benzene on aluminum:

1) 12.5; 2) 22.5; 3) 32.5; 4) 42.5°C.

Killmann and Schneider /99/ proved the reversibility of adsorption in the given case, and calculated the change in the enthalpy of adsorption for the systems studied (Table 8). The authors showed that at low concentrations the adsorption is an endothermal reaction, while at high concentrations it is an exothermal reaction. The authors explain these phenomena by the theory of monolayer adsorption, and assume that if in the first adsorption layer the molecules occupy a few active sites on the surface, then in the second layer they must displace segments of molecules adsorbed in the first layer, and also the solvent molecules. Therefore, the enthalpy decreases with occupation of the active sites of the layer.

TABLE 8. Enthalpy of adsorption in the poly (ethylene glycol)—aluminum—benzene system

A, mg	ΔH, kcal/mole	Range of concentrations, mg/ml
Molecular weight 28,000		
10	6.8	10—20
20	4.3	10—20
30	2.6	10—20
40	1.2	10—20
50	0	10—20
90	-4.9	20—50
100	-4.1	20—50
150	-1.1	20—50
Molecular weight 6000		
15	1.8	10—20
25	1.4	10—20
35	0.6	10—20
50	-4.0	20—50
70	-4.3	20—50
90	-3.8	20—50

It is our opinion that the polymer is not adsorbed layer-by-layer from concentrated solutions, but in molecular aggregates. Therefore, when analyzing the temperature dependence of adsorption, we must allow for the change in the character of structure formation in the solution under the effect of heat. It is clear that at higher concentrations, when aggregates are formed, the temperature dependence will be more marked; this is confirmed in /99/.

The above data show that the sign of the temperature coefficient of adsorption may change, depending on the properties of the system. The simultaneous effect of different factors must be taken into account. A temperature increase may lead to either an increase or a decrease in the solvent power, and since this parameter does not always affect the adsorption in the same way, the temperature dependence of the adsorption may be complex. The adsorbent — solvent interaction also changes with temperature.

Finally, when we consider the adsorption of polymers from concentrated solutions, we must take into account the effect of temperature on the size of the aggregates of molecules present in the solution and able to pass onto the

adsorbent surface. Thus, a temperature increase may lead to a growth
of the aggregates, but the solvent power will then deteriorate, and the
increase in the size of the aggregates will increase adsorption to a certain
limit only /58, 59/.

From the temperature dependence of the adsorption we can calculate
the thermal effect of the adsorption, which is very important for an
evaluation of the thermodynamic interaction between polymer and surface.

The "isosteric" heat of adsorption can be calculated from the Clausius –
Clapeyron equation /77/

$$\frac{\Delta H}{R} = \frac{H_0 - H_L}{R} = \frac{d \ln \frac{a_L}{a_0}}{d (1/T)}, \tag{3.7}$$

where H_0 and H_L are the partial molar enthalpies of the polymer on the
adsorbent surface and in the solution; a_0 and a_L are the corresponding
activities of the polymer. If we take $a_L = c$ and do not take into account
the temperature dependence of the activity of the macromolecules when
equal amounts are adsorbed, we have

$$\frac{\Delta H}{R} = \frac{d \ln c}{d (1/T)}. \tag{3.8}$$

Hence, if from the adsorption isotherm we take the values corresponding
at different temperature to equal amounts of the sorbed substance and plot
them on a graph according to the Clausius – Clapeyron equation, then the
slope of the straight line gives the enthalpy change when the polymer passes
from the bulk of the solution onto the surface. This method was used by
many authors to calculate the heat of adsorption, although in some cases
there was no evidence of the reversibility of the adsorption, while in the
Clausius – Clapeyron equation complete reversibility and equivalence of
the adsorption process are assumed.

We also applied this method to evaluate the heats of adsorption /201/.
We assumed that the time taken for adsorption equilibrium to be established
in the system is much shorter than the relaxation time of the slow
processes that take place in the polymer and which involve changes in the
degree of aggregation, structural rearrangement, etc. With reference to
such processes, adsorption can be considered to be a quasiequilibrium
process.

Our experimental data, and the data of other authors calculated from
(3.8), are summarized in Table 9. It can be seen that the heats of
adsorption are negative, that is, adsorption proceeds with an increase in
enthalpy, since the adsorption increases with temperature in the majority
of cases. However, this rule is not general, and for some systems we
observe the conventional character of adsorption, that is, heat is evolved.

It follows from the above data that the adsorption of polymers proceeds
with absorption of heat, and in this it differs from the adsorption of low-
molecular weight substances. For an explanation of this anomaly we must
bear in mind that the value of the heat of adsorption found includes the
intrinsic heat of adsorption, that is, the heat of interaction Q_{12} of the

polymer with the surface, the heat of desorption Q_{32} of the solvent molecules from the surface, and the heat of reaction Q_{13} of the polymer with the solvent. Thus, the overall heat of adsorption is

$$- \Delta H = q = Q_{12} - Q_{32} - Q_{13}. \qquad (3.9)$$

TABLE 9. Differential heat of adsorption from solutions on different adsorbents

A, g/g	Heat of adsorption, cal/mole	A, g/g	Heat of adsorption, cal/mole
Fiberglass—poly (methyl methacrylate) — acetone		Fiberglass—polystyrene— benzene	
0.02	-8910	0.007	-8580
0.04	-7260	0.010	-8780
0.08	-5280	0.015	-12,300
0.12	-3960	0.026	-16,500
Fiberglass—poly (methyl methacrylate)—toluene*		Fiberglass—polystyrene— cyclohexanone	
0.0015	-2020	0.02	-9500
0.0045	-8710	0.03	-9900
0.0060	-13,100	0.04	-9900
0.0075	-15,050	0.09	-9700
Fiberglass—poly (methyl methacrylate)—chloroform		Iron—poly (vinyl acetate)— carbon tetrachloride /77/	
0.010	-5360	0.0012	-6200
0.020	-6140	0.0013	-4900
0.026	-7130	0.0014	-4100
0.032	-7720	0.0015	-3700
		Copper — poly(vinyl acetate) — benzene	
		0.0025	-9050
		0.0050	-7020
		0.0075	-8770

* To the maximum on the adsorption isotherm.

The negative heat of reaction indicates that the last two terms of equation (3.9) are larger than the first. From equation (3.9) we can conclude that the heat of adsorption becomes more negative as the concentration of the solution increases, since in this case it is more difficult for the polymer molecules to pass onto the surface. However, as the adsorption increases with increase in concentration, the firm bond between the macromolecule and the surface becomes weaker, and the heat of adsorption decreases. The above data illustrate that q may increase or decrease, depending on the given system and the extent of adsorption.

The spontaneous adsorption process, involving an increase in the enthalpy of the system, should be accompanied by an entropy that is appreciably greater than the enthalpy, to ensure a negative value of the

free energy during adsorption. Koral et al. /77/ explained the negative
heat of adsorption by this fact. The entropy may increase because the
adsorption of the polymeric molecule on the surface leads to the transition
of a large number of solvent molecules from the surface into the bulk of
the solution. This must give a large excess entropy with reference to
the decrease in entropy as the result of bonding and restricting the mobility
of the macromolecules of the polymeric chain on the surface. Probably the
partial increase in the entropy during adsorption can be explained by the
packing of the polymeric macromolecule on the surfaces, which is less
close than in the bulk.

An analysis of the dependence of the heat of adsorption on surface
coverage shows that in adsorption from good solvents, such as polystyrene
from benzene, poly (methyl methacrylate) from chloroform, poly (vinyl
acetate) from carbon tetrachloride, q decreases with increase in this
parameter. However, in adsorption from poor solvents, q depends slightly
on surface coverage (polystyrene from cyclohexanone) or decreases
(poly (methyl methacrylate) from acetone, poly (vinyl acetate) from
benzene). We think that the stronger dependence of heat of adsorption
on the surface coverage in poor solvents is due to changes in the chain
conformation and conditions of structure formation in the solution when
the concentration is increased.

EFFECT OF MOLECULAR WEIGHT
ON ADSORPTION

The dependence of the amount of adsorbed polymer on its molecular
weight requires a detailed analysis, since from these data we can deduce
the structure of the adsorption layer, the number of adsorption centers,
and the bond energy per molecule. Usually the amount of polymer adsorbed
from the solution on establishment of equilibrium is determined
independently of whether during adsorption the molecular-weight distribution in the adsorbed film changes with reference to the molecular-weight
distribution of the polymer remaining in the solution. From the measurements, no conclusions can be drawn on the binding of molecules of a
certain size. However, when preferential or displacement adsorption is
studied, such conclusions can be made.

We have mentioned that the rate of adsorption of smaller molecules is
higher than that of larger molecules due to diffusion. Therefore, smaller
molecules are adsorbed from the polymer solution in the initial stage;
these are then displaced by larger molecules, corresponding to the thermodynamic equilibrium state.

However, if the adsorptional bond is very strong, such a displacement
may not occur, and the smaller molecules will be preferentially adsorbed.
Preferential adsorption of larger and smaller molecules is determined by
the porosity of the adsorbent.

We shall consider experimental data on the adsorption of polymers of
different molecular weights.

In some papers, an increase in the adsorption with molecular weight of
the polymer is reported. Brooks and Badyer /109/ found that the

adsorption of nitrocellulose on starch increases with increase in the molecular weight of the polymer. Poly(ethylene glycol) of low molecular weights (300 to 6000) is adsorbed more on carbon from aqueous solutions when the molecular weight increases /90/. The adsorption of low-molecular weight polyesters on glass and silica also increases with increase in molecular weight /73/. It is interesting to note that if the molecular weights are equal, the adsorption of the unsaturated polyester is four to five times that of the saturated compound /122/.

In the adsorption of polystyrene on chrome ferrotype plates (Figure 39) /62/ and of polyisobutylene and butyl rubber on carbon black (Figure 40) /84/, the adsorption increases with increase in molecular weight. During the adsorption of the copolymer of butadiene and styrene on carbon black, the adsorption increases up to MW = 500,000, and then it changes only slightly /107/. An increase in the molecular weight of poly(vinyl acetate) from 33,000 to 100,000 also increases its adsorption on iron powder and cellulose fibers /113/.

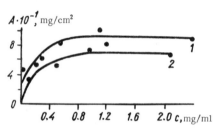

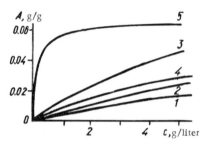

FIGURE 39. Adsorption isotherms of polystyrene with MW = $1.37 \cdot 10^6$ (1) and $0.54 \cdot 10^6$ (2) on chrome ferrotype plates.

FIGURE 40. Adsorption isotherms of polyisobutylene with MW = 5100 (1), 8300 (2), 14,000 (3), and of butyl rubber with MW = 8800 (4), 325,000 (5) on carbon black from n-hexane solutions.

Patat and Schliebener /88/ found that the adsorption of poly(methyl methacrylate) (MW = 44,000 to 1,500,000) increases appreciably with molecular weight. An increase in the adsorption of this polymer with molecular weight on quartz surfaces was observed by Sonntag and Jenkel /123/.

A linear relationship between adsorption and molecular weight was established in the adsorption of linear poly(dimethyl siloxane) on carbon black, titanium dioxide, and other pigments (Figure 41) /50/. A considerable increase in the adsorption of poly(dimethyl siloxane) on iron and glass powder with increase in molecular weight was observed by Perkel and Ullman /54/. Figure 42 shows that with increase in molecular weight the adsorption of poly(ethylene glycol) on aluminum from benzene increases. When the molecular weight of this polymer changes from 200 to 20,000, the adsorption on montmorillonite also increases (Figure 43) /124/.

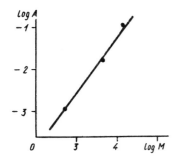

FIGURE 41. Adsorption of poly(dimethyl siloxane) as a function of molecular weight.

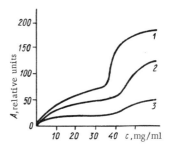

FIGURE 42. Adsorption isotherms of poly(ethylene glycol) with MW = 28,000 (1), 6000 (2), and 1000 (3) from benzene solutions on aluminum at 42.5°C.

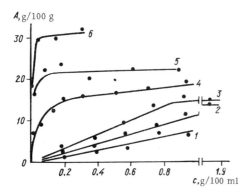

FIGURE 43. Adsorption isotherms of poly(ethylene glycol) with MW = 200 (1), 300 (2), 400 (3), 600 (4), 1500 (5), and 20,000 (6) on montmorillonite.

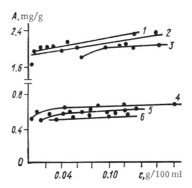

FIGURE 44. Adsorption of poly(methyl methacrylate) on glass (1—3) and iron (4—6) from benzene:

1,4) MW = $227 \cdot 10^3$; 2,5) $532 \cdot 10^3$; 3) $23 \cdot 10^3$; 6) $118 \cdot 10^3$.

The adsorption isotherms of poly(methyl methacrylate) on glass and iron from benzene solutions /112/ also indicate an increase in adsorption with increase in molecular weight (Figure 44). A similar dependence of the adsorption on molecular weight is observed in the ethylene-vinyl acetate copolymer — glass — benzene system /111/.

In a study of β-cyanoethyl esters of poly(vinyl alcohol) and of the β-chloroethyl ester of methacrylic acid on glass powder, Zakordonskii and Polonskii /119/ used the intrinsic viscosity [η] as the criterion for the molecular weight. They found that with increase in [η], that is, in the molecular weight, the adsorption of the polymers studied increases.

An analysis of the published experimental data shows that on smooth surfaces, on adsorbents with a small specific surface (iron, glass, etc.), and on porous adsorbents, the adsorption increases with molecular weight in the range of comparatively low molecular weights.

However, the opposite effect is also observed: a decrease in adsorption with increase in molecular weight. Thus, for the adsorption of poly(vinyl acetate), nitrocellulose, dextran, and synthetic rubber on carbon, Claesson /125/ found lower values for molecular weights exceeding 10,000. According to Frisch et al. /98/, this decrease is due to the short duration of the adsorption and the slow rate of desorption. Such a phenomenon may, of course, occur, as seen in Figure 11.

If the selected duration of the experiment is insufficient for establishing adsorption equilibrium, we may form incorrect conclusions as to the dependence of the adsorption on molecular weight because of differences in the diffusion rate of the molecules. However, we must also allow for the fact that when the molecular weights of the polymers adsorbed on porous adsorbents from solutions are high, the molecules cannot penetrate into the adsorbent pore. In this case a decrease in the molecular weight may lead to higher adsorption. This evidently explains the increase in the adsorption with decrease in molecular weight in the following systems: polystyrene — (methyl ethyl ketone + methanol) — carbon /74/; dextran — carbon black — benzene /126/; polyisoprene — benzene — carbon black /127/. It can be seen from the example of the adsorption of poly(methacrylic acid) on chromatographic alumina (Figure 45) /115/, or of poly(methyl acrylate) (PMA) and poly(methyl methacrylate) on alumina or carbon black DG-100 (Table 10) /128/ that the adsorption increases with decrease in molecular weight.

In some cases the adsorption is independent of the molecular weight, for example, in the system poly(vinyl acetate) — benzene — glass (Figure 46) /47/.

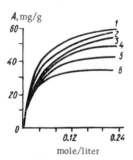

FIGURE 45. Adsorption isotherms of poly(methacrylic acid) on alumina:

1) MW = 57,000; 2) 67,900; 3) 111,000; 4) 240,000; 5) 447,000; 6) 1,170,000.

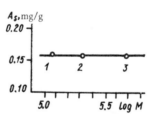

FIGURE 46. Adsorption as a function of molecular weight in system poly(vinyl acetate) — benzene — glass:

1) MW= 120,000; 2) 21,000; 3) 480,000.

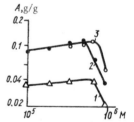

FIGURE 47. Adsorption as a function of molecular weight in the system polybutadiene — carbon black:

1) specific surface 99; 2) 108; 3) 125 m²/g.

A study of the adsorption of polymers on porous adsorbents over a fairly wide range of molecular weights indicates a complex change in adsorption with molecular weight. Thus, Kraus and Gruver /86/ showed that adsorption varies nonmonotonically with molecular weight in the system polybutadiene — carbon black (Figure 47).

TABLE 10. Adsorption and molecular weights of polymers

M · 10⁻⁶	A, mg/m²	
	on Al₂O₃	on carbon black DG-100
Poly (methyl acrylate)		
73.50	0.33	0.34
59.60	0.36	0.43
47.80	0.47	0.52
31.60	0.59	0.67
26.90	0.69	0.87
Poly (methyl methacrylate)		
0.60	0.17	—
0.36	0.18	0.16
0.20	0.19	0.26
0.11	0.41	0.46

The position of the peak on the adsorption — molecular weight curve is thus determined by the type of carbon black employed, which is apparently related to the pore size in the adsorbents used. Kraus and Gruver compared the pore sizes in carbon black and the size of polybutadiene molecules.

The mean pore radius was determined from the formula

$$r = 2V/S, \qquad\qquad (3.10)$$

where V is the weight of the carbon black; S is the specific surface. It was found that for carbon blacks with specific surfaces of 108, 125, and 99 m²/g, the mean pore size is 251, 316, and 619 Å, respectively. These magnitudes were compared with the molecular sizes of polybutadiene in heptane. The hydrodynamic volume v was calculated from

$$v = 100\,[\eta]\,M/N_A = 4\pi R^3/3, \qquad\qquad (3.11)$$

where N_A is the Avogadro number; R is the radius of gyration. For polybutadiene of molecular weight 500,000, $R = 360$ Å, that is, the magnitude has the same order as the pore dimensions in carbon black. The pore size in carbon black with a specific surface of 108 m²/g is somewhat smaller than the radius of gyration of the polybutadiene molecule. Therefore, for this carbon black, a decrease in adsorption begins after the polymer has attained a molecular weight that is lower than that required for other types of carbon black.

Howard and McConnel /84/ studied the adsorption of poly (ethylene oxide) on Aerosil silica, carbon, and nylon, and also found a complex dependence of the adsorption on the molecular weight (Figure 48). In the adsorption of this polymer on carbon, the dependence of the adsorption on molecular weight differed from that in the previous case (Figure 49). In the latter case the adsorption is almost independent of the molecular weight when MW = $10^3 - 2 \cdot 10^5$. An exception is adsorption from methanol. Howard and McConnel believe that this is due to the strong dependence of the solubility of poly(ethylene oxide) in methanol on the molecular weight. An even stronger dependence of the adsorption of this polymer from methanol on molecular weight is found on Aerosil silica (see Figure 34).

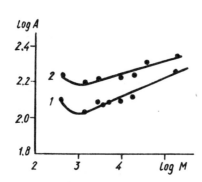

FIGURE 48. Graph of adsorption plotted against molecular weight in system poly(ethylene oxide) — benzene — Aerosil:

1) standard Aerosil; 2) Aerosil silica.

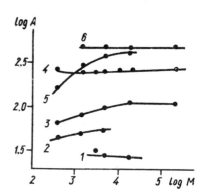

FIGURE 49. Graph of the maximum adsorption of poly(ethylene oxide) on charcoal at 25°C plotted against the molecular weight:

1) dimethylformamide; 2) chloroform; 3) dioxane; 4) benzene; 5) methanol; 6) water.

Howard and McConnel, like Kraus, estimated the size of the pores in carbon black, and found that the mean pore radius is 18 Å. They also studied the pore size distribution, and found that this distribution is wide, with a maximum at a radius of 22 Å. About half of the pores had a radius of 30 Å, and only slightly more than 10% of the adsorbent surface consisted of 60 Å pores.

Adsorption data were compared, and the size of the poly(ethylene oxide) molecules in benzene at 25°C was determined:

Molecular weight	1000	5000	10,000	20,000	50,000	100,000
$(S^2)^{1/2}$, Å	11	29	43	60	106	156

The results show that the size of the molecules penetrating into the adsorbent pores is smaller than the size estimated by the method of Flory, since the molecule is deformed when entering the pores. Moreover, the size of the polymeric conformations in the solution will depend on the solvent power.

For a further analysis of the results, information on the actual size of the molecules penetrating into the pores is necessary, but no such information was obtained by Howard et al. This makes it more difficult to obtain a reliable dependence of the adsorption on the molecular weight (Figure 49).

When the size distribution of the pores in nylon is compared with the adsorption of poly (ethylene oxide) of different molecular weights on this adsorbent, Howard et al. /84/ found a qualitative correspondence between these parameters (Figure 50).

The authors found that the size distribution of the pores in nylon is narrow, with a maximum frequency at a pore radius of about 15 Å. About 85% of the nylon surface consists of pores with a radius of 35 Å. It is therefore not surprising that the adsorption of poly (ethylene oxide) decreases with increase in molecular weight on this adsorbent. Most of the pores are inaccessible to the molecules of the polymer studied.

In the adsorption of polymers on nonporous surfaces, maximum adsorption A_S and molecular weight are related by

$$A_S = KM^{\alpha}, \qquad (3.12)$$

where K and α are constants.

From this equation we can estimate the conformation of the adsorbed polymeric chain on the surface of the solid body.

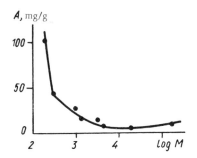

FIGURE 50. Graph of adsorption plotted against molecular weight in the system poly(ethylene oxide)–benzene–nylon.

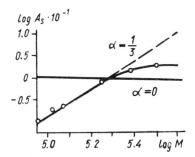

FIGURE 51. Graph of the adsorption plotted against molecular weight in the system copolymer of ethylene and vinyl acetate–benzene–glass.

The graph of the adsorption plotted against molecular weight for the system copolymer of ethylene and vinyl acetate − glass − benzene (Figure 51) /111/ shows that $0 < \alpha < \frac{1}{3}$. The glass surface can be considered to be a nonporous adsorbent, and therefore it can be assumed that the adsorbed molecules are attached to the glass surface in the form of loops.

Perkel and Ullman /54/ found that the dependence of the adsorption of poly (dimethyl siloxanes) on the molecular weight is described by

$$A_S = K_1 + K_2 M^{\alpha}. \qquad (3.13)$$

The values of K and α are given in Table 11. The constants vary with the nature of the solvent and surface within the limits $0.2 < \alpha < 0.5$.

TABLE 11. Values of K and α for the adsorption of poly (methyl siloxane) and poly (methyl methacrylate)

Surface	Solvent	$K \cdot 10^2$	α
	Poly (methyl siloxane)		
Glass	Benzene	0.97	0.40
	Hexane	2.94	0.35
Iron	Benzene	0.34	0.43
	Hexane	4.90	0.23
	Poly (methyl methacrylate)		
Iron	Benzene	0.35	0.04
Glass	Benzene	2.00	0
Iron	1,2-dichloro-ethane	0.14	0.08

In the adsorption of poly (methyl methacrylate) on different surfaces /112/, α is close to zero. This indicates a planar configuration of the macromolecules on solid surfaces. In the system poly (vinyl acetate) — glass —benzene, α is also equal to zero. Hence, the adsorbed molecules are distributed over the plane of the adsorbent. For the system poly (dimethyl siloxane) — titanium dioxide, $\alpha = 1$, and the molecules are distributed in the adsorbed film as bristles. The ideas on the conformation of the adsorbed chains will be explained in more detail in Chapter 4.

According to Koral et al. /77/, $\alpha = 0.0 - 0.2$ in the absorption of poly (vinyl acetate) from various solvents; α is smaller in a good solvent than in a poor one. Patat et al. /88/ showed that α decreases with increase in temperature.

All the values of α vary between the values for the two theoretically possible limiting cases of adsorption. Thus, the macromolecules are adsorbed as loops or balls. Such an analysis of the structure of the adsorption film is, of course, only possible in adsorption from dilute solutions. In adsorption from concentrated solutions which do not contain isolated molecules but their aggregates, the adopted models are inapplicable.

Some authors have discussed the preferential adsorption of molecules with a high or low molecular weight. In most cases conclusions are formed from data on the variation in the viscosity of the solutions over the adsorbent. According to Kolthoff and Gutmacher /91/, the intrinsic viscosity decreases in the adsorption of the copolymer of styrene and butadiene on carbon black. The authors consider that this is evidence of the preferential adsorption of large macromolecules.

According to the data of /91/, the viscosity first begins to increase, then decreases, and reaches its final value within 48 hours. The amount of polymer adsorbed becomes constant even before the constant viscosity is reached. If a polymer of lower molecular weight is added to the solution the viscosity remains constant, but if a high-molecular weight fraction is added, the viscosity decreases appreciably. The authors consider that these results are a convincing proof of the preferential adsorption of large molecules. El'tekov /72/ studied the adsorption of polystyrene on carbon black, and measured the intrinsic viscosity of the supernatant liquid. He also concludes that larger molecules are preferentially adsorbed in the equilibrium state.

This opinion is also shared by Felter et al. /53/, who studied the molecular weight distribution of poly(vinyl chloride) in adsorption. Gel permeation chromatography was used for this research. Calcium carbonate (Purecal U) was the adsorbent, and the solvent was chlorobenzene. Figure 52 shows the curves characterizing the MWD of the adsorbed polymer (curve 1) and the polymer in solution (curve 2) after $19\frac{1}{4}$ hr of adsorption. The area under curve 1 represents the weight fraction of the adsorbed polymer, and the area under curve 2 the weight fraction of the polymer remaining in solution. Table 12 shows the results of the determination of the molecular weight in the adsorption process. The number-average $\overline{M}_{n_a}$ and the weight-average $\overline{M}_w$ molecular weights in the adsorbed film were calculated from

$$\overline{M}_{w_0} = f_p \overline{M}_{w_p} + f_a \overline{M}_{w_a}, \tag{3.14}$$

$$\frac{1}{\overline{M}_{n_0}} = \frac{f_p}{\overline{M}_{n_p}} + \frac{f_a}{\overline{M}_{n_a}}, \tag{3.15}$$

where f_p and f_a are the weight fractions of the polymer in the solution after adsorption and in the adsorbed film, while $\overline{M}_n$ and $\overline{M}_w$ are the number-average and weight-average molecular weights. The subscripts 0, a, and p refer to the initial polymer in solution, the adsorbed polymer, and the polymer after adsorption. Table 12 shows that $\overline{M}_{w_a}$ and $\overline{M}_{n_a}$ are much higher than $\overline{M}_{w_p}$ and $\overline{M}_{n_p}$. Thus, Felter et al. concluded that there is preferential adsorption of the high-molecular weight fractions of poly(vinyl chloride) on Purecal.

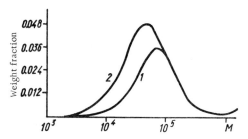

FIGURE 52. Molecular-weight distribution of the adsorbed polymer (1) and the polymer in solution (2) in the system poly(vinyl chloride)—Purecal—chlorobenzene.

Miller and Pacsu /129/ fractionated cellulose acetate on porous carbon, and found that the smaller molecules are preferentially adsorbed, and are not replaced later by larger ones.

TABLE 12. Variation in molecular weights during adsorption in the system poly (vinyl chloride)—Purecal—chlorobenzene

Time of adsorption, hr	$\bar{M}_{n_p} \cdot 10^{-4}$	$\bar{M}_{n_a} \cdot 10^{-4}$	$\bar{M}_{w_p} \cdot 10^{-4}$	$\bar{M}_{w_a} \cdot 10^{-4}$
0	—	—	—	—
0.25	3.18	5.33	6.75	9.17
1.50	3.16	5.06	6.63	9.27
3.20	3.09	5.39	6.63	9.22
19.25	3.13	4.84	6.81	8.66

Remark. $\bar{M}_{n_0}$ and $\bar{M}_{w_0}$ are equal to $3.65 \cdot 10^4$ and $7.65 \cdot 10^4$, respectively.

Si Jung Ye and Frisch /96/ chromatographically eluted a mixture of polystyrene of different molecular weights on a column filled with activated charcoal. The polystyrene mixture was introduced into the column and then eluted by the solvent. When the solvent (methyl ethyl ketone) in which the polystyrene was dissolved was used, the polymer could not be eluted. With toluene, a better solvent, it was possible to elute low-molecular weight polystyrene. The best solvent of those studied, namely, tetralin, eluted 77% of the polymer. The fractionation may be due to the weaker bonds of the low-molecular weight fractions with the solid surface, or to the more rapid dissolution of the low-molecular weight fractions in the solvent, or to both these phenomena. However, no unequivocal conclusion on the adsorption of molecules of low or high molecular weight can be made.

The interpretation of such tests becomes even more complicated if we also take into account the effects of pore structure and the irreversibility of adsorption. Thus, in studies on the preferential adsorption of low-molecular weight polymers /127/ and in fractionation /85/ kinetic effects due to the pore size evidently played a certain part. When they studied the adsorption of poly (methyl siloxane) on iron and glass, Perkel and Ullman /54/ found a preferential adsorption of the low-molecular weight fraction, although the amount adsorbed increases with increase in molecular weight. Since the system is appreciably irreversible, preferential adsorption is due to the fact that the low-molecular weight fraction migrates to the surface more rapidly, is irreversibly sorbed on it, and is not displaced by the high-molecular weight fraction. It is possible that the preferential adsorption of the low-molecular weight fractions of cellulose acetate /129/ and poly (vinyl acetate) /89/ on carbon is due to the irreversibility of the adsorption.

Conclusions on the preferential adsorption of macromolecules of different molecular weight are usually based on measurements of the viscosity of the solution above the adsorbent.

This was studied in detail in the adsorption of two different fractions of poly (vinyl acetate) on glass /130/. The experimentally found dependence of the viscosity on surface coverage was compared with the theoretical dependence. For the theoretical consideration, four cases were taken: 1) only the high-molecular weight fraction is adsorbed: 2) only the low-molecular weight fraction is adsorbed; 3) an equivalent number of molecules of both fractions is adsorbed; 4) fractions of the same weight are adsorbed. The authors analyze the dependence of $[\eta]$ on the surface coverage Θ for a solution containing two fractions with molecular weights M_1 and M_2.

If only the high-molecular weight fraction is absorbed (case 1), then

$$[\eta]_{m_1} = K\left[M_1^\alpha - \frac{N_2 M_2 \, (M_1^\alpha - M_2^\alpha) \, 10^2}{(N_1 M_1 + N_2 M_2) \, 10^2 - A_{S_1} \omega N_A \Theta}\right], \qquad (3.16)$$

where α is the constant in (3.13); N_1 and N_2 are the number of molecules of molecular weight M_1 and M_2 in the fractions; A_{S_1} is the maximum adsorption of the polymer; ω is the weight of the adsorbent; N_A is the Avogadro number. For case 2, the following equation was derived:

$$[\eta]_{m_2} = K\left[M_2^\alpha + \frac{N_1 M_1 \, (M_1^\alpha - M_2^\alpha) \, 10^2}{(N_1 M_1 + N_2 M_2) \, 10^2 - A_{S_2} \omega N_A \Theta}\right]. \qquad (3.17)$$

When an equivalent number of molecules n_1 and n_2 of both fractions is adsorbed (case 3), we have the equation

$$[\eta]_n = k\,\frac{(N_2 - n_1)\,M_1^{\alpha+1} + (N_2 - n_1)\,M_2^{\alpha+1}}{(N_1 - n_1)\,M_1 + (N_2 - n_2)\,M_2}. \qquad (3.18)$$

For case 4, the authors derived the equation

$$[\eta]_w = K\left\{\frac{M_1^\alpha - M_2^\alpha}{2} + \frac{(N_1 M_1 - N_2 M_2)\,(M_1^\alpha - M_2^\alpha)\,10^2}{2\,(N_1 M_1 + N_2 M_2)10^2 - 2A_{SW}N_A\Theta}\right\}. \qquad (3.19)$$

Some of the above equations were applied to experimental results on the adsorption of poly(vinyl acetate) on glass /130/. Two cases were tested, viz., cases 4 and 3. The experimental and theoretical curves of $[\eta]$ plotted against Θ were compared, and it was found that, for equivalent weight concentrations, $[\eta]$ decreases with increase in adsorption. This comparison shows that a decrease in $[\eta]$ does not necessarily indicate the preferential adsorption of high-molecular weight fractions. In adsorption from a mixture containing an equivalent number of molecules of poly(vinyl acetate) of two fractions, $[\eta]$ is little dependent on Θ. Hence, there is no preferential adsorption of low- or high-molecular weight fractions. In this system, the polymer is adsorbed independently of the molecular weight, that is, the adsorption is not fractional, and only at a certain coverage the adsorption of the low-molecular weight fraction increases because of the higher rate of diffusion.

Thus, from an analysis of the dependence of $[\eta]$ on the degree of surface coverage, it can be seen that $[\eta]$ cannot be used to indicate the preferential adsorption of a fraction, since even when there is equal (with reference to number of molecules) adsorption of the high- and low-molecular weight fractions, $[\eta]$ will change because of the difference in the contribution of these fractions to the intrinsic viscosity.

EFFECT OF NATURE OF ADSORBENT
ON ADSORPTION

Experimental data on the adsorption of polymers from solutions show that the degree of adsorption of the same high-molecular-weight fraction varies over a wide range, depending on the nature of the adsorbent.

Brooks and Badyer /109/ found that in the adsorption of nitrocellulose from a cyclohexane — acetone mixture on corn starch, a greater amount is adsorbed on this adsorbent than on potato starch. The authors attribute this to the difference in the specific surface of the adsorbent. Kraus and Dugone /76/ obtained similar results in the adsorption of the copolymer of styrene and butadiene on carbon blacks with different specific surfaces. They showed that the adsorption increases with increase in the specific surface of the black. Increase in the specific surface of cellulose sorbents also leads to an increase in the adsorption of poly(vinyl acetate) from ethyl acetate solutions /113/.

Adsorbents with similar specific surfaces but differing in their chemical nature have different adsorptive capacities.

Figure 53 shows the adsorption isotherms of one of the polyesters studied by Stromberg /73/ from chloroform solutions on different adsorbents. The lowest adsorption is observed on glass, which is the adsorbent with the smallest specific surface ($0.67 \ m^2/g$). The specific surfaces of alumina and silica are the same ($200 \ m^2/g$), but the adsorption of polymers on these adsorbents differs. Hence, the chemical nature of the surface affects adsorption. Stromberg believes that the positively charged ions present on a glass surface lead to adsorption because bonds are formed with the oxygen of the ester group of the polymer.

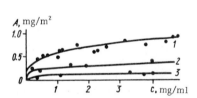

FIGURE 53. Adsorption of poly (neopentyl succinate) on glass (1), silica (2) and alumina (3) from chloroform at 30°C.

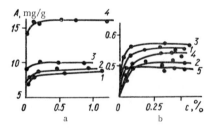

FIGURE 54. Adsorption isotherms of the polymers of methacrylic acid esters on Al_2O_3 (a) and glass (b):

1) poly (methyl methacrylate); 2) poly (ethyl methacrylate); 3) poly (propyl methacrylate); 4) poly (butyl methacrylate); 5) poly (amyl methacrylate).

However, Jenkel and Rumbach /110/ studied the adsorption of poly (methyl methacrylate), and observed an increase in the adsorption in the sequence carbon, quartz, glass, alumina. A considerable increase in the adsorption of methacrylic acid esters on passing from glass to alumina was also observed by other authors /132/ (Figure 54).

No appreciable differences were observed in the adsorption of some polymers on smooth surfaces of different chemical nature. Thus, Killman et al. /99/ did not observe any significant changes in the adsorption of polymers on metal surfaces (aluminum, platinum, copper). However, the adsorption is higher on cellulose than on metals. The authors explain this by the presence of hydroxyl groups on the cellophane surface.

The equilibrium adsorption of poly(ethylene glycols) on the smooth surfaces of glass and aluminum /99/ is practically equal, but during the formation of the first step on the isotherms (we have already noted that a staggered isotherm is characteristic of this process because of the multi-layer adsorption of the polymer), the adsorption on glass is much stronger than on aluminum.

We should note that a comparison of the results of different investigators apparently working with adsorbents of the same type is very difficult, since the methods of preparation of the specific surface may differ, and this greatly affects adsorption.

Some investigators note that water attached to the adsorbent surface frequently affects adsorption. Perkel and Ullman /54/ studied the adsorption of poly(dimethyl siloxane) from a benzene solution on glass, and found that if the adsorbent is heated in vacuo at 300°C, adsorption of the given polymers increases. Heat treatment of the glass leads to removal of the OH group from its surface and to the formation of

$$-\overset{|}{\underset{|}{Si}}-O-\overset{|}{\underset{|}{Si}}-$$ bonds. The adsorption of poly(dimethyl siloxane) on the

original surface of the glass can be represented by

Poly(dimethyl siloxane) is adsorbed on the treated surface according to the scheme

Evidently, the type of bond shown in scheme II is more effective than that in scheme I, and thus the adsorption of poly(dimethyl siloxane) is greater on heat-treated glass.

Stromberg and Kline /73/ studied the effect of the heat treatment of glass and silica on the adsorption of polymers, and showed that treatment of the glass does not change adsorption (Figure 55a). This is apparently because the tubes with heat-treated glass were unsoldered in air, and only then filled with the polymer solution. The adsorbent probably adsorbed water from the air. Perkel and Ullman /54/ filled the tubes in an atmosphere of dry nitrogen, and thus prevented the entrance of water vapor from the air into the adsorbent. In a later paper Stromberg /49/ reports on the considerable effect of the heat treatment of glass on the adsorption of polyesters.

Heat treatment of silica (Figure 55b) strongly decreases the adsorption of polymers. The adsorption in the system polystyrene – activated carbon is similarly changed if the adsorbent is dried /98/. Here the presence of traces of water increases adsorption, apparently because of the differing mechanisms of adsorption interaction in the given systems.

Heat treatment of Aerosil silica and partial elimination of hydroxyl groups from its surface reduces the adsorption of the epoxide resin ED-5 (Figure 56) /118/. In this case it can be assumed that the adsorption of the macromolecules of epoxide resin is specific in character, probably because of the negatively charged epoxy groups $CH_2 - CH_2$, which can

react specifically with the protonated hydrogen atoms of the hydrophilic surface groups.

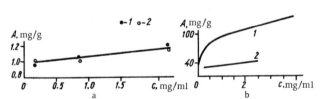

FIGURE 55. Adsorption of poly(ethylene adipate) from toluene on glass (a) and of polyethylene-o-phthalate on silica (b) at 30°C:

1) original adsorbent; 2) heat-treated adsorbent.

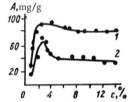

FIGURE 56. Adsorption isotherms of ED-6 from solutions in toluene on Aerosil silica dried at 170 (1), and ignited at 550°C (2).

Kiselev et al. /46, 133, 134, 135/ systematically studied the effect of the chemical nature of the adsorbent surface on the adsorption of polymers from solutions. They examined the effect of modification of the surface of highly dispersed solid solutions and the effect of the chemical nature of the filler particles on the character of adsorption. Figure 57 shows the adsorption of poly (neopentyl phthalate) from heptane on different adsorbents /133/. It should first be noted that heat treatment of Aerosil silica at 700°C decreases adsorption by 20%. The specific surface of the Aerosil silica after heat treatment remains practically constant. The authors believe that the adsorption is due to the partial elimination of the hydroxyl groups from the Aerosil silica surface by this treatment.

The adsorption of the polymer by the large-pored silica gel S-41 with a hydroxylated surface is almost 25% of the adsorption of the same polymer by the original Aerosil silica. This is the result of the difference in the specific surface of these adsorbents (for Aerosil silica $S_{sp} = 170$ m^2/g, for S-41, $S_{sp} = 41$ m^2/g). Treatment of the Aerosil surface by an organo-silicon compound sharply reduces the adsorption.

When Kiselev et al. /133/ processed the results, they converted the adsorption to grams per unit surface. However, incorrect conclusions can be formed on the effect of the nature of the surface and of the thickness of the adsorption film, since not all the surface determined according to the adsorption of small molecules is accessible to large polymeric molecules.

We must also take into account the aggregation of the filler particles in the polymeric solution, as this reduces the effective adsorbent surface.

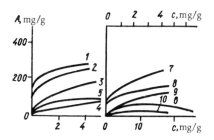

FIGURE 57. Adsorption isotherms of poly (neo-pentyl phthalate) from heptane solution on Aerosil silica treated at 180 (1) and 700°C (2) on Al$_2$O$_3$ (3), silica gel ShSM (4), highly porous silica gel S-41 (5), and Aerosil silica modified by reaction with (CH$_3$)$_3$SiCl (6), on the following carbon blacks: highly dispersed (7), original Ukhta (8), and graphitized (9), and on TiO$_2$ (10) (the scale for 7−9 is shown at the top).

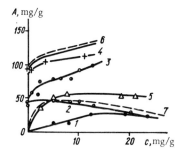

FIGURE 58. Adsorption isotherms of poly (dimethyl siloxane) from heptane solutions on different adsorbents:

1) channel black; 2) carbon black oxidized by H$_2$O$_2$; 3) carbon black oxidized by HNO$_3$ + H$_2$SO$_4$; 4) silica gel BS-280; 5) Aerosil silica modified by acrylonitrile; 6) Aerosil silica hydroxylated and partially dehydroxy-lated; 7) Aerosil silica modified by trimethylchlorosilane.

Figure 58 shows the adsorption isotherms of poly (dimethyl siloxane) (MW = 350,000) when the character of the Aerosil silica and carbon black surfaces is changed because of thermal, mechanical, or chemical treatment /133, 134/. The figure shows that the temperature of Aerosil silica treatment has practically no effect on the adsorption of poly (dimethyl siloxane), in contrast to the adsorption of poly (neopentyl phthalate). It appears that after heat treatment at 900°C, the amount of OH groups on the Aerosil silica is still sufficient for the formation of hydrogen bonds with the terminal OH groups of the poly (dimethyl siloxane) macromolecules. If the OH groups on the Aerosil silica surface are replaced by trimethyl-silyl or acrylonitrile groups, there is less possibility of hydrogen bridge formation with the remaining silanol groups, and dispersion interaction in the polymer — filler system is reduced because of the formation of a fairly thick layer of organosilicon or organic substances /134/.

Similar results indicating a decrease in the adsorption of poly (dimethyl siloxane) after the adsorbent (silica gel BS-280) had been treated by trimethylchlorosilane were obtained by Trapeznikov et al. /136/. Treatment of Aerosil silica by trimethylchlorosilane and diazomethane also reduces the adsorption of poly (ethylene oxide) /83/.*

The strong adsorption of poly (dimethyl siloxane) from solutions in hexane on the hydroxylated surface (see Figure 58) is due to the presence of hydroxyl groups on these macromolecules /134/. Kiselev et al. believe that the slight adsorption of this polymer on channel black is due to the strong adsorption of the solvent molecules on this adsorbent and also to the roughness of the adsorbent surface, which favors the adsorption of small molecules /137/. The increase in the adsorption of poly (dimethyl siloxane) on oxidized carbon blacks is explained by the decrease in the adsorption of hexane molecules, which is confirmed by the adsorption of hexane vapors on the original and oxidized carbon blacks. Moreover, when channel black is oxidized by the acid mixture, a large number of carboxyl groups (up to 28 mg-eq/g) are formed. They react specifically with the terminal groups of poly (dimethyl siloxane), and reduce the nonspecific (dispersion) interaction of the nonpolar solvent (n-hexane) with the surface of carbon black.

As in the case with poly (dimethyl siloxane), the adsorption of polystyrene decreases when Aerosil silica treated by organosilicon compounds is used as the adsorbent (Table 13). Hence, modification of the filler by organo-silicon compounds, in the same way as their thermal treatment, usually leads to a decrease in the adsorption of polymers. However, sometimes we also observe some increase in adsorption when the surface of the filler is modified. El'tekov et al. /72/ note that in the adsorption of polystyrene from toluene or benzene solutions, the adsorption of polymers increases when the surface of Aerosil silica is heat treated. The authors state that the removal of a considerable amount of hydroxyl groups from the surface of the filler leads to a decrease in the adsorption of the solvents (benzene and toluene), the molecules of which react with these groups.

TABLE 13. Peak adsorption of polystyrene on Aerosil silica modified in various ways

Aerosil silica	A, mg/m^2 at C_0 = 3 mg/g	
	MW = 40,000	MW = 290,000
Original	0.90	1.10
Treated by trimethylchlorosilane	0.70	0.60
Treated by methyldichlorosilylmethyl propionate	0.55	0.35
Heated at 90°C	—	1.20

* [Reference incorrect.]

Thus, if the surface of adsorbents is heated, there is a large increase in the adsorption of polymers from solutions that have molecules that specifically react with hydroxyl groups (alcohols, ketones, esters, aromatic hydrocarbons). However, removal of the hydroxyl groups from the surface may lead to reduced adsorption from more inert solvents (for example, saturated hydrocarbons).

Figure 59 shows that the value of the adsorption of polymers on mineral adsorbents (macroporous silica gel, Aerosil silica, kaolins) are very similar /138/. Appreciable differences are observed in the adsorption of polyacrylamide on graphitized carbon black and silica. Unfortunately, no conclusions on the effect of the chemical nature of the surface on the adsorption can be made in this case, since the surfaces of the adsorbents were treated in different ways. Additional research is required on the effect of the chemical nature of the surface on the adsorption of polymers to obtain conclusive evidence. Nevertheless, the results of papers /75, 79, 133—135/ clearly show that the chemical nature of the adsorbent, especially modification of its surface (thermal, chemical), strongly affects the adsorption of polymers from solutions.

Different types of surfactants are frequently employed as filler modifiers. Apparently, this modification of filler will affect the adsorption of polymers, and hence the properties of the filled molecules. These problems are discussed in detail by Tolstaya et al. /139—145/.

These authors studied the effects of the degree of filler modification by various surfactants /144, 145/. The graph of the adsorption of perchloro-vinyl resin from toluene solutions plotted against the degree of modification of Prussian blue modified by various surfactants (Figure 60a) shows that covering the surface of the filler by stearic acid and octadecylamine (which form a chemical bond with Prussian blue) leads to a decrease in adsorption /143/. However, octadecyl alcohol is physically sorbed on the surface of the given filler, and the Prussian blue modified by it does not change the adsorption of perchlorovinyl resin.

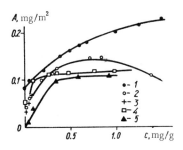

FIGURE 59. Adsorption of polyacryl-amide (MW = 30,000) from aqueous solutions on channel black treated at 3000°C (1), macroporous silica gel (2), Aerosil silica (3), and kaolins (4,5).

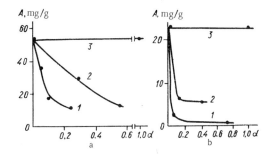

FIGURE 60. Graph of the adsorption of perchlorovinyl resin from toluene solutions plotted against the degree of surface coverage of Prussian blue (a) and rutile (b):

1) stearic acid; 2) octadecylene; 3) octadecyl alcohol.

Similar results on the effect of modification of the filler by surfactants were obtained when studying resin PFV-3, which is an alkyd resin based on pentaerythrite and phthalic anhydride. The effect of modifying the surface by stearic acid is stronger than that by octadecylamine. The difference is even greater in the adsorption of perchlorovinyl resin on modified rutile (Figure 60b). This difference between the action of the two surfactants is the result of the adsorption of the acid and the resin on similar centers of the pigment surface, and the fact that the degree of surface coverage of rutile by stearic acid is higher than by octadecylamine /143/.

Spectral studies /140, 142/ showed that the spectrum of the rutile surface covered by stearic acid remains almost unchanged after the adsorption of perchlorovinyl resin. This confirms the results of adsorption measurements, which indicate the absence of appreciable adsorption of the resin on rutile, with maximum coverage by stearic acid. In the spectrum of rutile, preliminarily modified to saturation by octadecylamine and after adsorption of the resin, a weak absorption band of the resin appears at 1426 cm^{-1}. This indicates insignificant adsorption of the resin on the surface. In adsorption of the resin on rutile treated with octadecyl alcohol, the spectrum of the rutile changes appreciably (Figure 61). The 1470 cm^{-1} absorption band characteristic of the alcohol disappears, and the intense absorption band of the resin appears. This is due to direct adsorptional changes, indicating displacement of the alcohol from the rutile surface.

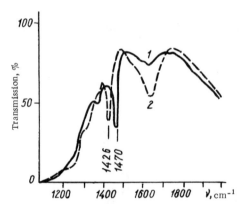

FIGURE 61. Infrared spectra of rutile modified by octadecyl alcohol (1), and after adsorption of per-chlorovinyl resin on it (2).

Modification of the filler surface by surfactants thus reduces the adsorption of the polymer when the agents are chemisorbed but does not change the adsorption of polymers when the surfactants are physically adsorbed on the filler.

However, sometimes a more complex dependence of the adsorption of polymers on the modification of the surface of the filler is observed. This was proved by Tolstaya et al. / 139, 143/ when they studied adsorption in the system rutile – dichlorostearic acid – perchlorovinyl resin – dichloroethane. The curve of the dependence of the adsorption of the resin on the degree of surface coverage of the rutile by the surfactant (Figure 62) has a peak. At this peak the surface coverage of rutile is 0.12, that is, the coverage is incomplete. The bond between the modifier and filler is strong, as shown by spectral studies (Figure 63). In the infrared spectrum the characteristic adsorption bands of the resin (1426 cm^{-1}) and of the salts of dichlorostearic acid (1630 cm^{-1}) are observed; these do not disappear from the spectrum of the washed adsorbent. Modification of the rutile surface by octadecylamine does not lead to adsorption of the polymer on the pigment (curve 2).

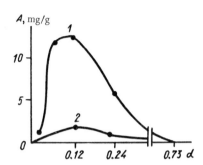

FIGURE 62. Curve of the adsorption of per-chlorovinyl resin on rutile from dichloro-ethane plotted against the degree of surface coverage of the adsorbent by dichlorostearic acid (1) and octadecylamine (2).

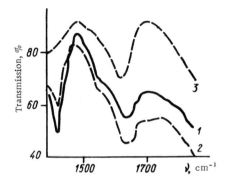

FIGURE 63. Infrared spectra of rutile at incom-plete surface coverage by dichlorostearic acid (0.2%) and adsorption of perchlorovinyl resin from dichloroethane (1), as (1) + washing of the pigment by solvent (2), and infrared spectrum of the original rutile (3).

From these results Tolstaya et al. /139, 143/ concluded that only an adsorption film of the surfactant with molecular properties similar to those of the polymer initiates polymer adsorption when the surface coverage is incomplete. However, the mechanism of this initiation is still unclear. When the surface coverage of the film by dichlorostearic acid is complete, the adsorption of the polymer does not take place (or decreases greatly), as in the case of other surfactants.

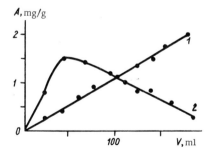

FIGURE 64. Adsorption of perchlorovinyl resin (1) and octadecylamine (2) on rutile when these substances simultaneously interact with the pigment.

A study of the adsorption of perchlorovinyl resin from the mixture resin — octadecylamine on rutile /143/ showed that the amine is preferentially adsorbed on this adsorbent, but the adsorption of the resin decreases sharply (Figure 64).

Thus, not only the chemical nature of the adsorbent, but also the modification of adsorbents of the same nature, lead to changes in the adsorption of polymers over a wide range.

ADSORPTION OF POLYMER MIXTURES

Mixtures of several polymers or oligomers are used in the production of filled polymers and coatings. Therefore, the study of the adsorption of mixtures of polymers is important. However, very little has been done in this field. Apparently there are difficulties in determining the concentrational changes of the individual polymers used in the mixture.

Thies /80/ studied the simultaneous adsorption of polystyrene and poly (methyl methacrylate) on silicas from dilute trichloroethylene solutions at 25°C. The concentration of the solution was determined from the characteristic bands of polystyrene (697 cm^{-1}) and poly (methyl methacrylate) (1720 cm^{-1}). A constant amount of the polymers was used in the mixture, while the amount of adsorbent was varied all the time. The isotherms of the mixtures were the curves of the dependence of the equilibrium concentration c_0 of both polymers on the adsorbent weight ω. The fractions of the segments on the adsorbed polymeric chain p bound to the adsorbent surface were calculated from the shift in the characteristic bands, with allowance for the concentration of the adsorbed copolymer (c_a) and the measured concentration of the bound segments (c_b).

Preliminary experiments on the adsorption of individual substances showed that in this case the usual Langmuir isotherms are obtained. The value of p for poly (methyl methacrylate) with MW = $3.68 \cdot 10^5$ increased from 0.29 to 0.35. Larger values of p indicate strong compression of the adsorbed film. The p values for polystyrene with MW = $2.46 \cdot 10^5$ are half those for poly (methyl methacrylate).

Typical adsorption isotherms of polymer mixtures are represented in Figure 65. The solid line shows the calculated curve of c_0 vs ω for poly (methyl methacrylate), while the dotted line shows this curve for polystyrene. The calculation was carried out with two assumptions: 1) poly (methyl methacrylate) is completely adsorbed before the adsorption of polystyrene begins, and polystyrene does not interfere with the adsorption of poly (methyl methacrylate); 2) polystyrene begins to be adsorbed immediately the adsorption of poly (methyl methacrylate) is finished, and the presence of the latter does not hinder the adsorption of polystyrene.

Figure 65 shows that assumption (1) is valid. However, the measured values of c_0 for polystyrene are somewhat larger than the calculated ones, although the adsorption of polystyrene is almost independent of the presence of poly (methyl methacrylate) on the adsorbent surface.

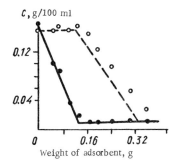

c, g/100 ml

0.12

0.04

0   0.16   0.32

Weight of adsorbent, g

FIGURE 65. Adsorption isotherm of
a mixture of polystyrene and poly-
(methyl methacrylate).

These results indicate the preferential adsorption of poly (methyl methacrylate) under equilibrium conditions on the silica surface because of the stronger interaction of the polymer with the surface of the given adsorbent.

Experiments were carried out on the preliminary adsorption of different quantities of polystyrene on the adsorbent surface, followed by the addition of excess poly (methyl methacrylate) into the adsorption system. It was found that poly (methyl methacrylate) rapidly and completely substitutes polystyrene on the silica surface. The degree and rate of substitution are practically independent of the surface coverage of the adsorbent by poly- styrene, and of the holding time of the polystyrene — adsorbent system. We know that with variation in these parameters the conformation of the adsorbed molecules may change. It is clear that in this case the substitution of the molecules of one polymer for the molecules of another does not lead to changes in surface of the adsorbent.

The complete substitution of poly (methyl methacrylate) for polystyrene on the silica surface indicates that the adsorption of polystyrene is reversible, and confirms the hypothesis of some reseacch workers /77, 112/ on the preferential adsorption of surfactants. It is worth noting that poly (methyl methacrylate) is substituted for polystyrene, although these polymers are incompatible.

All the experiments were carried out under conditions far from phase separation. It was assumed that the poly (methyl methacrylate) molecules rapidly penetrate to the silica surface through the adsorbed polystyrene film. The incompatibility of the polymers does not affect the process. There is only a slight difference in the *p* values of the adsorbed polystyrene and poly (methyl methacrylate) molecules when these molecules are adsorbed from a mixture or adsorbed from individual substances. This again indicates the constant conformation of the adsorbed macromolecules.

Thies /147/ studied the effect of nonpolymeric adsorbates on the adsorption of poly (vinyl acetate) and the copolymers of vinyl acetate and ethyleme on silica from different solvents. Methanol, acetonitrile, cyclo- hexanol, 1,8,9-anthracenetriol, and dioxane were the nonpolymeric adsorbates. They were introduced in amounts of 5—10% into the solvent (trichloroethane or cyclohexane). It was found that these additives decrease polymer adsorption, and hence the fraction of carbonyl groups directly bound to the surface.

Infrared spectroscopic studies showed that the additives react with the carbonyl of the ester group to form complexes, which also affect polymer — surface interaction.

Bothman and Thies /148/ studied more concentrated mixtures (1.5— 5.5 wt%) in which the concentration of polymer was higher than the concentration at which the system separates into two phases /148/, in contrast to the case when the system was far from the point of two-phase separation /80/. The sorbent was mixed with the two-phase system, and the adsorption from the phases where the concentrations of the components differed proceeded in different ways.

We studied the adsorption of the mixtures poly (vinyl acetate) + poly (ethylenevinyl acetate), poly (dimethyl siloxane) + poly (dimethyl siloxane) with a small fraction of vinylic side groups, and ethylcellulose + polystyrene on silica. We determined the adsorption isotherms and the fraction of the segments of the adsorbed macromolecules for the diverse mixtures. We also studied the displacement of one polymer from the surface after the introduction of another polymer.

We found that the preferential adsorption of one of the components may increase under the effect of the other component, incompatible with the first one. The adsorption is changed most when the solution separates into two phases. Because of the incompatibility of the components, an increase in the adsorption of one component is accompanied by an increase in the fraction of its segments p bound to the surface. From data on the displacement of one polymer by another, we established the sequence of affinity of the polymers to silica (adsorption from trichloroethane): PVA > ethylcellulose > PMMA > poly (ethylenevinyl acetate) > polystyrene.

Schick and Harvey /150/ studied the adsorption of a mixture of a polar with a nonpolar polymer on the carbon black Graphon – liquid interface. Two patterns of competitive adsorption have been studied, viz., sequential competitive adsorption (displacement of an initially adsorbed polymer by another), and simultaneous competitive adsorption (the solid surface is exposed to both polymers simultaneously). In the former case the displacement of the molecules of one polymer by the molecules of another polymer can be observed. The following polymers were used: nonpolar ^{14}C labeled polystyrene (PST), polar poly (vinyl acetate) (PVA), and polar poly (methyl methacrylate) (PMMA). Toluene with a high solvent power for polystyrene, and butan-2-one, a poor solvent for this polymer, were used.

First PVA was adsorbed on carbon black from a solution containing the polymer in a concentration of 0.8 mg/ml. The PST solution was then added, and its adsorption on Graphon with the preliminarily adsorbed PVA was studied (Figure 66). In the good solvent the adsorbed PVA practically did not change the adsorption of PST, while in the poor solvent the adsorption of PST decreases under these conditions. The authors of /150/ believe that this pattern confirms the hypothesis that the weakly adsorbed PVA can be completely displaced if the PST molecules are adsorbed on the carbon black from the good solvent as a flat oriented nonlayer. Conversely, PVA hinders the formation of a thick layer of adsorbed PST molecules, formed in adsorption from the poor solvent, when only some of the molecules are directly bound to the surface.

Thus, in a medium of a solvent poor for PST, PVA does not completely replace PST on the Graphon surface. However, the nonpolar PST is preferentially adsorbed to a high degree on the nonpolar adsorbent. The effect of the solvent power plays a larger role for the nonpolar polymer than for the polar polymer.

In the simultaneous adsorption of two polymers, the pattern is more or less independent of the way the adsorption is studied (Figure 67). With increase in the amount of PVA introduced into its mixture with PST, the adsorption did not change.

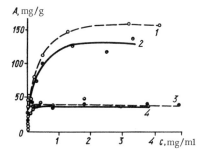

FIGURE 66. Displacement of polystyrene by poly (vinyl acetate) from Graphon at 25°C:

1) adsorption of pure PST from butan-2-one on Graphon; 2) adsorption of PST on Graphon which had previously adsorbed PVA; 3) adsorption of pure PST from toluene; 4) the same after adsorption of PVA on Graphon.

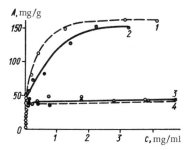

FIGURE 67. Simultaneous competitive adsorption of polystyrene and poly (vinyl acetate) on Graphon at 25°C:

1) pure PST in butan-2-one; 2) PST + PVA (1 : 1) in butan-2-one, PST + PVA (1 : 2); PST + PVA (1 : 3); 3) pure PST in toluene; 4) PST + PVS in toluene.

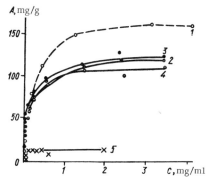

FIGURE 68. Simultaneous competitive adsorption in the presence of polymethacrylates and the copolymer of vinyl acetate and ethylene on Graphon from butan-2-one solution at 25°C:

1) pure polystyrene; 2) PST + PMMA, MW = = 388,000 (1 : 1); 3) PST + poly (isobutyl methacrylate), MW = 410,000 (1 : 1); 4) PST + PMMA, MW = 416,000 (1 : 1); 5) PST + + copolymer (1 : 1).

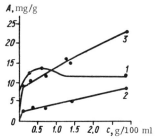

FIGURE 69. Isotherms of the adsorption on rutile of resin AS-4 (1), copolymer of vinyl chloride and vinyl acetate (2) from chloroform, and of resin AS-4 from a solution of its mixture with the copolymer in chloroform (3).

In the adsorption of polystyrene in the presence of polymethacrylates and the copolymer of vinyl acetate and ethylene, the polymethacrylates reduce the adsorption more than PVA (Figure 68). The authors believe that this can be explained by the greater strength of the bond between methacrylates (compared with PVA) and Graphon, or by the difference in the conformation of the molecules of the polar polymers on Graphon. The nature of the polymethacrylates slightly affects the adsorption of polystyrene, while the effect of the molecular weight of the polymethacrylate is stronger. The adsorption of polystyrene decreases in the presence of the copolymer of vinyl acetate and ethylene, which contains both a polar and nonpolar moiety.

The results indicate the preferential adsorption of the nonpolar polymer in its mixture with the polar one on the surface of the nonpolar adsorbent.

Uvarov et al. /151/ consider that the application of IR spectroscopy to research on the adsorption of polymer mixtures is insufficient, since in some cases the adsorption bands of the polymers are near one another. The authors recommend the simultaneous application of UV and IR spectroscopies. They applied this method to a study of the adsorption of the copolymer of vinyl chloride and vinyl acetate (87 : 13), and the amide-acrylic resin AS-4 (glycerophosphate modified by castor and sunflower oils and copolymerized with a mixture of methyl and butyl esters of methacrylic acid) on titanium dioxide from chloroform.

The principle of the method is as follows. First the calibration curves showing the concentrational dependence of the optical density of the adsorption band of the $C=O$ group of each of the polymers in the $1700-1800$ cm^{-1} region are plotted. At the same time the solutions are analyzed on an SF-4 spectrophotometer in the UV region ($250-300$ mμ). This region contains the intense absorption peak of the resin AS-4 (275 mμ), while the copolymer in the concentrations studied yields no such peak. After the concentration of the resin AS-4 in the mixture has been determined, the level of the copolymer in the solution is calculated from the adsorption band of the $C=O$ group in the infrared region of the spectrum, with allowance for the known concentration of the resin AS-4.

Figure 69 shows that only the resin AS-4 is adsorbed from the copolymer mixture. Apparently, the molecules of this resin form a stronger bond with the adsorbent surface than the molecules of the other copolymer. It is also interesting that the resin is adsorbed much better from a mixture than from a solution of the resin only. This confirms the fact that preferential adsorption of one of the components may increase under the effect of the other component, incompatible with the first /148/.

Thus, from an analysis of the data at present available on the adsorption of a polymer mixture, it is not possible to draw any general conclusions on regularities in this process. Further research in this direction is necessary.

Chapter 4

STRUCTURE OF THE ADSORBED FILM AND
CONFORMATION OF THE ADSORBED CHAINS

Most isotherms of the adsorption of polymers from dilute solutions are
curves that reach saturation at certain concentrations. In principle, such
curves are characteristic of Langmuir monomolecular adsorption.
However, the thicknesses of the adsorbed film calculated from these isotherms
are much greater than those observed for monomolecular adsorption
according to the classical concept. This is due to the specific features
of the adsorption of polymers, the structure of the adsorbed films, and the
conformation of the polymeric chains adsorbed on the surface.

To solve the problem of the disagreement between the adsorption
isotherms and the film thicknesses calculated from them, we must study
in detail the problems involved in the structure of the adsorbed film. This
problem is also important for understanding the mechanisms of the
processes occurring on the interface with the solid in filled and reinforced
polymer systems and other heterogeneous polymeric materials. The
adsorption of the polymer is, of course, the first stage of the formation of
the bond between polymer and solid surface on glueing, application of coating, stabilization of dispersions, etc.

The simplest hypothesis on the structure of the adsorbed film is the
following: the polymeric molecules rest flatly on the surface and form
a large number of Van-der-Waals bonds with the surface. However, such
a distribution of molecules on the surface, even if possible in principle,
in the initial stages of adsorption with a small surface coverage, cannot be
used to describe the equilibrium film, since the experimentally determined
thickness of the adsorbed film appreciably exceeds the thickness of the
macromolecules.

Therefore, several structural models of the adsorbed film were
suggested. Jenkel and Rumbach /110/ assume that the polymeric molecule
adsorbed by the surface forms wrinkles or loops on it, stretching from
the surface into the polymer solutions. These wrinkles form a characteristic "bristle" on the surface. Only the segments of the chains which are at
the end of the loops take part in the adsorption bond, while the other
segments do not directly interact with the adsorbent, and are bound to it
only via other segments.

Such a hypothesis on the structure of the film explains why its thickness
exceeds that of a monolayer. In this case the structure will depend on the
strength of the segment — surface bond. If the bond is strong, a small
number of the segments will be bound, since the stronger adsorption bond

will hinder transition of the molecules from the surface into the solution. However, if the interaction is weak, a larger number of bonds (per molecule) is required to retain the molecules on the surface, and the size of the wrinkles or loops will be smaller. For their calculation, Jenkel and Rumbach /110/ assumed that the density of the polymer in the adsorbed film is the same as in the block. However, this assumption is not justified, and introduces considerable errors into the calculation of the layer thickness. The authors also proved that the inner surfaces are inaccessible to macromolecules, which introduces additional errors into the calculation.

According to Koral et al. /77/, the macromolecules adsorbed on the surface must have a statistically coiled formation (Gaussian chain) characteristic of the macromolecules in solution. They also assume that the polymer molecules are bound to the adsorbent surface at several sites, and that the larger part of the polymeric chain is not bound to the adsorbent surface but moves freely in the solvent near the surface to which the chain is bound. The area occupied by the polymeric molecule on the adsorbent surface can be calculated from the radius of gyration of the molecule in the solution. If the radius and the molecular weight of the polymer are known, a theoretical limiting adsorption A_∞ on saturation can be calculated.

Let n molecules with radius of gyration R accurately cover the whole adsorbent surface

$$S = n\pi R^2. \tag{4.1}$$

Then

$$A_\infty = 1000MS/\pi R^2 N_A, \tag{4.2}$$

where N_A is the Avogadro number; A_∞ is the amount of adsorbed polymer, mg/g; M is the molecular weight.

For poly(vinyl acetate), the molecular weight and the radius of gyration were calculated from light-scattering data in methyl ethyl ketone /77/. The calculation showed that the experimental values of adsorption are 7—40 times the values calculated by using the proposed model. This indicates that the model of the undistorted polymeric coil is too simple, and must be modified.

It can be assumed that the configuration of the molecule in the adsorbed film is distorted, so that it occupies a smaller area on the surface than it does in the solution. The molecule may have the shape of an elongated ellipse with the major axis perpendicular to the adsorbent surface. In this case, because of lateral compression, the entropy of the coil decreases. This must be compensated by excess enthalpy during adsorption. Koral et al. /77/ consider that such mutual entanglement and interpenetration of the coils are possible at high local concentrations. However, the role of distortion of the coil dimensions or the mutual entanglement of the macromolecules is not yet clear.

Further ideas on the structure of the adsorbed film were advanced by Patat et al. /88/ in papers on the adsorption of polystyrene and

polyvinylpyrrolidone of different molecular weights on various sorbents.
Their ideas are also based on the disagreement between the experimental
adsorption values and the values calculated from the hypothesis on a
monolayer. The authors also consider the model of the adsorbed film in
the form of "bristles" and in the form of "coils." In the first case data on
the interatomic distances make it possible to calculate the area occupied
by the segment of the molecule, and from this result to calculate the
adsorption and the theoretical weight of the adsorbed film. In the second
case the area occupied by the macromolecule is given by $F_0 = (\bar{h}_w/2)^2\, \pi$.
To form a monomolecular layer of area A_0, $N_m = A_0/F_0$ macromolecules
are required. Hence, the weight of the layer is

$$M_S = \frac{N_m\bar{M}_w}{L} = \frac{A_0\bar{M}_w}{F_0L} = \frac{4A\bar{M}_w}{\bar{h}_w^2 L\pi},\qquad(4.3)$$

where L is the Loschmidt number; M_w is the molecular weight. The
number of adsorbed films can be determined from

$$Z_S = M_P/M_S = M_P F_0 L/A_0 M_w,\qquad(4.4)$$

where M_P is the experimental amount by weight of the polymer adsorbed
by unit surface. The film thickness is found from

$$D_S = \bar{h}_w Z_S = \frac{\bar{h}_w M_P F_0 L}{A_0 M_N} = \frac{M_P L \bar{h}_w^3}{4A_0\bar{M}_w}.\qquad(4.5)$$

A comparison with experimental data shows that the measured amounts
of the adsorbed polymer cannot be placed into a single layer, even if it
is assumed that the molecules form a bristle on the surface, which gives
the closest packing (the packing is loosest when the surface is covered
by coils).

The authors of /88/ therefore assume multilayer adsorption, in which
changes in the shape of the macromolecules are not compulsory. This
hypothesis is confirmed by the fact that the adsorbed film must contain
solvent bound by solvation of the polymeric molecule. The authors believe
that the adsorbed film may contain polymeric molecules which are not
attached to the adsorbent surface even by a single molecule. This altogether
eliminates adsorption, even of deformed macromolecular coils. Thus,
there remains only multilayer adsorption, during which the molecules
interpenetrate and become entangled.

No papers have yet appeared in which molecular interaction of the
polymeric chains in the adsorbed film and its role in the structure of the
adsorbed film are theoretically considered. Lipatov, Sergeeva, and their
co-workers developed entirely different ideas, according to which
macromolecular aggregates, and not single macromolecules, pass onto
the adsorbent surface. The aggregates are formed in the solutions, even
at relatively low concentrations /37, 38, 161/. Such an approach leads
to another investigation on the structure of the adsorbed film, from the
point of view of structure formation in the polymer solution. However,
the authors did not calculate the thickness of the films because no
quantitative data on structure formation in solutions were available.

Fundamental ideas on the structure of the adsorption boundary layer were advanced in the papers of Jenkel, Rumbach and Patat et al., which we have already discussed. All the later papers are based on the idea that the macromolecules are attached to the surface only, via a small number of chain segments, while the remaining segments remain in solution and do not directly interact with the surface. Such ideas may agree with both the adsorption model of loops on the surface and the adsorption model of macromolecular coils. The authors of all further studies developed these ideas and attempted to find a theoretical justification for them.

First, the number of macromolecular segments that interact directly with the adsorbent surface had to be found. Fontana and Thomas /64/ were the first to solve this task. They used infrared spectroscopy to determine the ratio between the number of segments attached to the surface and the total number of segments p in the poly (methyl methacrylate) — silica system. Some of their data on the surface coverage of the adsorbent (Θ) are shown in Table 14. It can be seen that about 36% of the segments are attached to the surface when the surface coverage is high. Such a p value indicates that the molecule is present on the surface in a rather elongated state, which corresponds to the Simha — Frisch — Eirich theory /173/ for systems with a strong reaction between polymer and adsorbent.

TABLE 14. Determination of the fraction of adsorbed segments

Polymer—solvent	A_s, mg/g	Θ	Amount of SiO_2, mg/cm²	Adsorbed segments, mg/cm²		Total adsorbed polymer	Fraction of adsorbed carbonyl groups
				from C=O data	from OH data		
Poly (lauryl methacrylate)	314	0.95	0.67	0.073	0.073	0.210	0.35
(MW = 330,000)—	234	0.71	1.28	0.107	0.105	0.298	0.36
dodecane	—	—	1.23	0.102	0.110	0.289	0.35
	120	0.36	1.29	0.069	0.068	0.155	0.44
	—	—	1.44	0.077	0.067	0.173	0.44
Poly (lauryl methacrylate)	52	0.16	1.57	0.032	0.012	0.082	0.39
(MW = 3,330,000)—	272	0.96	1.03	0.100	0.115	0.272	0.37
decalin	181	0.64	0.97	0.081	0.089	0.176	0.46
	—	—	1.04	0.079	0.088	0.188	0.42
Poly (lauryl methacrylate)	292	0.96	0.97	0.105	0.125	0.283	0.37
(MW =1,190,000)—	216	0.71	1.28	0.103	0.116	0.278	0.37
dodecane	101	0.33	1.10	0.059	0.064	0.111	0.53

These results also show that p tends to increase at low surface coverages. However, this tendency is weak, and we may assume that in this case the weak dependence of p on Θ is due to the fact that the configuration of the adsorbed chain is determined mainly by intramolecular effects, and not by molecular interaction of the adsorbed chains. In any case, under the experimental conditions adopted, the intermolecular forces never become appreciable.

   No dependence of p on the nature of the solvent from which the polymer
was adsorbed could be established. This indicates that the size of the
molecular coil has no influence on the degree of attachment of the
molecules to the surface. The degree of attachment of poly (methyl
methacrylate) and polystyrene to the surface of silica when adsorbed from
dilute trichloroethylene solutions was studied by IR spectroscopy /152/.

   The experimental values of p were compared with adsorption on different
sections of the adsorption isotherms. It was found that p depends on the
surface coverage. This indicates differences in the conformation of the
adsorbed molecules. For poly (methyl methacrylate), p is as high as 0.3,
which the author attributes to the formation of sufficiently compressed films
on the adsorbent surface. For polystyrene, p is considerably lower,
because of the smaller surface. The tendency of p to decrease at higher
degrees of surface coverage indicates the formation of a looser structure
of the adsorbed film.

TABLE 15. Attachment of active groups to surface in the adsorption of the vinyl acetate—ethylene
copolymer

Degree of surface coverage	Fraction of attached carbonyl groups	Fraction of attached copolymer segments
Poly (vinyl acetate)		
0.23	0.54	—
0.35	0.58	—
0.33	0.51	—
0.39	0.45	—
0.63	0.49	—
0.95	0.30	—
Vinyl acetate-ethylene copolymer (EVA-1)		
0.31	0.77	0.16
0.46	0.80	0.16
—	0.61	0.13
—	0.68	0.14
0.67	0.61	0.13
—	0.64	0.13
0.70	0.59	0.12
1.00	0.52	0.11
—	0.53	0.11
Vinyl acetate-ethylene copolymer (EVA-2)		
0.12	0.85	0.38
—	0.87	0.38
0.20	0.76	0.34
—	0.80	0.35
0.38	0.68	0.30
—	0.73	0.32
0.74	0.58	0.26
0.77	0.56	0.25
1.00	0.53	0.23

Table 15 shows the attachment of the active groups of the surface in the adsorption of the copolymer of vinyl acetate and ethylene. It can be seen that *p* depends on the level of the polar CO groups in the copolymer /153/. In this case *p* also depends on the surface coverage. When this parameter is low, *p* is 0.85, which indicates a strongly compressed adsorbed film. As the surface coverage increases, the structure of the adsorbed film continuously changes; this is expressed by changes in *p* and, judging from the *p* values, involves transition from flat to more extended molecules (Figure 70).

Peyser et al. /65/ studied the adsorption of poly (ethylene terephthalate) on silica surfaces, and compared their results with data on the adsorption of polystyrene. They found that the behavior of the polar polymer differs appreciably from that of the nonpolar. For polyesters with molecular weights between 5400 and 2500 the *p* values are 0.34 and 0.37, irrespective of the adsorption. The authors think that this is due to the fact that the conformation of the adsorbed molecule remains the same with increase in the concentration of the polymer in solution and increase in the degree of surface coverage. The relatively high *p* values are attributed to the planar distribution of the molecules in the adsorbed film.

The adsorbent particles had sizes comparable to those of the macro-molecules (0.015−0.020 μ). In this case it can be assumed that one molecule may be attached to several particles of the adsorbent. This also determines the stronger attachment to rough surfaces than to smooth. From a comparison with the data for polystyrene, it is possible to explain the differences in their behavior by the difference in the flexibility of the molecule and in the energy of interaction of the macromolecules with the surface. The rigidity of the poly (ethylene terephthalate) chain is the reason for the constant *p* values.

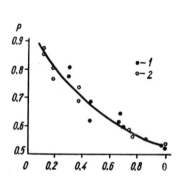

FIGURE 70. Graph of the attachment of carbonyl groups plotted against the degree of surface coverage θ :

1) EVA-1; 2) EVA-2.

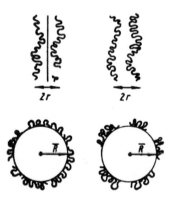

FIGURE 71. Variation in the hydro-dynamic dimensions as the result of adsorption on capillary walls or particles.

These results indicate the above-mentioned contradictions on the dependence of **p** on surface coverage, concentration of the solution, and other factors. The amount of experimental data is not yet sufficient to derive general relationships. However, the method for determining **p** appears to be very promising, since it can be used to explain the conformation of the adsorbed macromolecules. Such evaluations are of course qualitative only, since no theoretical correlation could be established between **p** and any parameter characterizing the conformation of the adsorbed macromolecules. The assumptions on a compressed film or a layer of extended molecules formed from **p** values are relative in character. A comparison of the experimental **p** values with those prediced by theory may give valuable information on the film structure.

In most of the published experimental studies, the problems involved in the configuration of molecules are considered, but not a single one reports a direct method for determining the conformation. (In principle, such methods could be developed by using polymers with labeled terminal groups so that it would be possible to estimate the distance between the chain ends.) Therefore, all conclusions on the conformation are obtained either from the fraction of attached segments, or from data on the film thickness and a comparison of the thickness with the size of the dissolved macromolecules. In many papers the thickness of the adsorbed polymer film is determined by different physical methods. The method of ellipsometry has been the most widely applied, as it can be used to determine the thickness of the adsorbed film and the polymer concentration in it, since the adsorbed film contains the bound solvent.

The problem of the thickness of the adsorbed film is mainly solved from data on the interfacial area and the amount of adsorbed polymer. It is assumed that the density of the polymer in the block is equal to that in the film. By using this assumption, the following thicknesses of the adsorbed film were found for different polymers and adsorbents /39/:

Polymer	Film thickness, $\overset{\circ}{A}$
Poly (vinyl acetate)	1500
Poly (ethylene glycol)	2400
Poly (methyl methacrylate)	7200
Poly (vinyl pyrrolidone)	7500
Poly (vinyl alcohol)	12,000

These results agree with viscometric results obtained by Fendler et al. /155/.

Kiselev and El'tekov /75, 79, 133/ developed a hypothesis on the complete unrolling of the polymeric chain on the adsorbent surface during strong specific adsorption, that is, spontaneous transition of the coil into a spiral. However, the specificity of the adsorption must be proved, and this is not possible for the systems studied by these authors (for example, polystyrene — graphitized carbon black). Therefore, the reported calculations of the thickness of the adsorbed film of polystyrene on rutile according to the equation $h = A_{max}/\rho$ (ρ is the density of the adsorbed polymer, taken as equal to the bulk density) cannot be theoretically justified, even if the thickness is of the order of 8 Å.

We should note that the data of Kiselev and El'tekov do not agree with the results of all the other papers. We believe that this is due to the calculation of adsorption per unit adsorbent surface, which is incorrect, as we have mentioned above.

Rowland et al. /157/ applied a new method for determining the film structure   In this, the effective thickness of the film is measured in the capillaries or on the surface of the disperse particles. Figure 71 shows the scheme for measuring the hydrodynamic dimensions of the disperse particles or the dimension of the capillary. From the results obtained for several systems, the authors unequivocally concluded that the polymer is adsorbed from the dilute solution in the form of monolayers, formed by the molecular random coils with a size proportional to that of the coils in the solution. The authors used Einstein's law and calculated the increase in the size of the disperse particles in solution. From these data and the interfacial area, the corresponding film thicknesses were computed. Figure 72 shows data on the variation in the viscosity of the suspension and increase in the size of the disperse particles as the result of adsorption. The mean size of the particles in the bulk was calculated from

$$d_o = \left(\sum n_i d_i^3 / \sum n_i \right)^{1/3}, \qquad (4.6)$$

where n_i is the number of particles of size d_i.

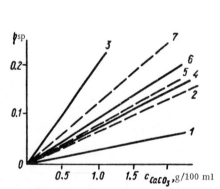

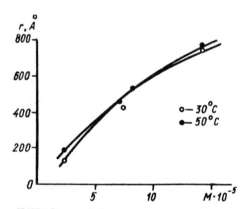

FIGURE 72. Dependence of the specific viscosity of a suspension of calcium carbonate in tetra-chloroethylene on the concentration:

1) according to Einstein's law for rigid spheres;
2) for completely dispersed $CaCO_3$ particles;
3) for unstabilized dispersion; 4—7) for dispersion with 0.2% PMMA of increasing molecular weight.

FIGURE 73. Graph of the film thickness of poly-(methyl methacrylate) adsorbed from benzene plotted against molecular weight.

To prevent flocculation of the disperse phase, which might distort the results, the viscosities of the suspensions extrapolated to infinite shear rate were used to determine the effective particle volume /68/. The authors studied the structure of alkyd resin on titanium dioxide, and also the adsorption of random copolymers of lauryl methacrylate, methyl methacrylate, glycidyl methacrylate, and others. The film thicknesses obtained were 100−240 Å. Such small thicknesses are evidently obtained because the experimental interfacial area of the adsorbent was used for the calculation.

Rowland and Eirich studied the thickness and structure of the adsorbed PMMA, PVAc, and PST films by measuring the flow rates of polymer solutions and pure solutions through porous Pyrex disks /158/. Figure 73 shows a typical dependence of the film thickness on the molecular weight calculated from these data. Very compressed films, in which the molecules strongly interact with the surface, should be characterized by a weak dependence of the thickness on molecular weight, while the dependence on the nature of the polymer should be retained. However, the experimental film thickness for polar and nonpolar polymers differs only slightly. Therefore, the authors assumed that it is not the polymer − surface interaction, but the polymer − solvent one (which is the dominating factor in the conformation of the adsorbed chain) that has the strongest effect on the film structure. In this case the film thickness and the intrinsic viscosity should be correlated (Figure 74). Such a linear correlation actually exists for polar polymers, which corresponds to data on variations in Δr (according to the increase in the volume of the disperse particles) /159/.

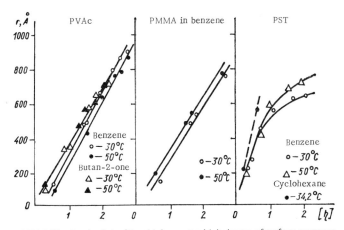

FIGURE 74. Graph of the film thickness at a high degree of surface coverage plotted against molecular weight.

Rowland and Eirich assume that the coil becomes deformed or compressed during adsorption /158/. A measure of such deformation is the ratio of the film thickness to twice the radius of gyration of the free coil in the solution, which is termed reduced thickness of the adsorbed film Δr^*. For a coil which is adsorbed without distortion, Δr^* should be about 0.9 (Figure 75).

Somewhat higher values are obtained for PST, indicating that the PST molecules are almost undeformed on the surface, and hence sorbed via a small number of segments. With increase in molecular weight the solubility decreases, the coil is compressed, Δr^* decreases, and the number of sorbed segments grows. Another pattern is observed for PMMA and PVAc, which is determined by their polarity, and leads to stronger adsorption. The authors believe that increase in molecular weight restricts the number of chain conformations on the surface, so that the number of loops formed increases continuously.

It is important that the amount of adsorbed polymer is not simply equivalent to the monolayer filled with the nondeformed coils. The area occupied by a sphere of radius r in two-dimensional hexagonal packing is 3.5 r^2, while the radius of gyration of a random walk coil is $1/_6 Nl^2$, where l^2 is an overestimate of the area of the segment.

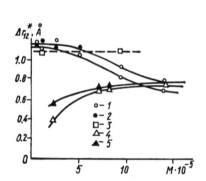

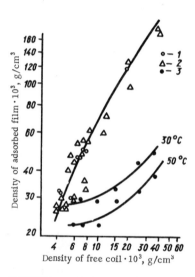

FIGURE 75. Graph of the reduced film thickness of a polymer at a high surface coverage plotted against molecular weight:

1) PST in benzene, 30°C; 2) PST in benzene, 50°C; 3) PST in cyclohexane, 34.2°C; 4) PMMA in benzene, 30°C; 5) PMMA in benzene, 50°C.

FIGURE 76. Graph of relative density of adsorbed film plotted against density of coil:

1) PMMA in benzene; 2) PVA in benzene and butan-2-one; 3) PST in toluene.

The monolayer filled with coils should therefore be equivalent with reference to the amount of adsorbed polymer to the monolayer consisting of segments lying flat on the surface. But the amount of adsorbed polymer is equivalent to 2—8 monolayers. Thus, the coils should interpenetrate or be subjected to strong lateral compression, which is very probable because of the low density of the segments (Figure 76). To reduce the density to

the value corresponding to the free coil, the polymer should occupy a
surface three times that available for nitrogen adsorption.

Figure 77 shows the graph of the film thickness plotted against the
amount of adsorbed polymer in various solvents. Two nearly horizontal
straight lines (a) show the effect of solvent power on r for PST-950,000
and PST-110,000 while the sloping lines show the trend with changing
molecular weight. As the solvent power decreases, the amount of the
polymer increases, while the thickness changes only slightly. The film
consequently becomes denser.

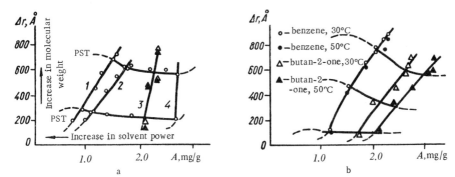

FIGURE 77. Graph of film thickness plotted against the amount of adsorbed PST, PMMA (a), and PVAc (b)
in various solvents:

1) PST in benzene at 50°C; 2) PST in benzene at 30°C; 3) PMMA; 4) PST in a Θ-solvent.

From these data, the following conclusions can be formed. The polymeric molecules are adsorbed as somewhat compressed or interpenetrating
coils. This follows from the shape of the isotherm, film thickness and
from the fact that the Δr — equilibrium concentration curves reach limiting
values much faster than the isotherms. At the critical molecular weight
loops begin to form, which stretch into the solution. The loops are formed
because of the increasing elastic resistance to compression, which in turn
increases because the relative deformation of the coil when it is drawn into
the surface increases with increase in molecular weight. In thermodynamic
terms, this means that with increase in molecular weight the loss in
conformational entropy because of the "flattening out" of the chains on the
interface begins to balance ΔH adsorption, and thus leads to increased
swelling of the adsorbed film. When the affinity to the surface is small
(for example, PST in a good solvent), the loops begin to form even at low
molecular weight.

Bulas /159/ proposed a very interesting hypothesis on the relative
density distribution of the segments in the film. The film consists of two
layers: a) a denser lower layer close to or on the surface, with a density
lower than that of the monolayer, consisting of segments, because of steric
hindrances; b) a more distant and less dense layer consisting of loops,
that is, of the polymer not directly attached to the interface. This layer
is responsible for the "apparent excess" adsorption. The contribution of
the loops to the effective hydrodynamic film thickness depends on the

volume concentration of the segments and the rigidity of the loops. However, the approximate proportionality of the intrinsic viscosity indicates that the upper layer has a structure similar to that of the free coils in solution. The fact that the solvent and temperature effects are the same in the adsorbed film and the solution phase confirms this pattern.

The frequently observed increase in adsorption with temperature can be explained not only by the positive enthalpy and positive entropy of adsorption, which are the results of the desorption of a large number of small molecules, but also by a lower surface coverage by segments and redistribution of segments over the surface, which leads to an increase in the number of active sites and an increase in adsorption.

Malinskii et al. /160/ calculated the thickness of the film adsorbed on the surface of disperse particles by viscometry. The authors studied the viscosity of solutions and melts of guttapercha and of solutions of isotactic polystyrene. The materials were filled with corundum powder (low polydispersity, mean size 5 μ) and glass powder (3.6 and 10 μ). The curves obtained could be described by the Einstein equation. The film thickness as a function of the concentration of the solution changed from 1 to 8 μ. The authors concluded, as we had previously (18, 161/, that molecular aggregates already existing in solution are adsorbed on the solid surface. The film thicknesses, which are larger than those published in other papers, are explained by the fact that the studies were carried out in concentrated solutions, that is, when, according to Lipatov /18/, structure formation in solutions must be taken into account.

The most interesting results on the structure and thickness of the adsorbed film are nevertheless obtained by ellipsometry, since by this method we can observe the time-dependent parameters of the film structure during adsorption. Stromberg et al. /62/ used ellipsometry in situ for investigating the relationship between the film thickness and the molecular weight, and also between the film thickness and the size of the macromolecules at the Θ-temperature for polystyrene sorbed on a chrome surface. For all molecular weights of the polymer, the authors observed an initial increase in the film thickness until a plateau is reached, again indicating a continuous increase in concentration. The concentration of the solution at which the plateau region is reached decreases with increase in molecular weight. In the plateau region the film thickness increases with increase in molecular weight.

To explain the variations in film thickness, the configuration of the macromolecules in dilute solutions is considered. Stromberg et al. assume that as the concentration of the solution changes, competition between the approaching polymeric molecules for sites on the surface begins. Hence, the conformation of the adsorbed chain should be subjected to continuous changes.

The conformation of the adsorbed molecule will thus depend on surface coverage, which agrees with the ideas developed from the p value according to IR-spectroscopic data. The process rate of adsorption studied by ellipsometry depends on two factors: a) arrival of the molecule from the solution at the surface and its attachment by at least one segment (this stage is determined by the concentration and by the free surface); b) the subsequent attachment of other segments of the adsorption chain, if it is attached by at least one segment, that is, "spreading out" of the molecule on the surface, which also depends on the amount of free surface available.

An analysis of the experimental time dependence of the film thickness shows that the initial rate of process (b) is much faster than the rate of process (a). As the surface becomes covered, competition for the active sites on the surface becomes greater, and if the first molecules could have a relatively flat conformation, the following molecules should be more extended. Finally, equilibrium conformation of the macromolecules is established, which depends on the molecular weight of the polymer and the concentration of the solution.

It is assumed that molecules adsorbed at different times have different conformations. The increase in the equilibrium film thickness with increase in concentration indicates that the conformation of polymeric chains depends on the concentration of the solution. In a more concentrated solution, the adsorbed molecule forms more loops, the number of points at which the molecule is attached decreases, while the layer thickness increases until the maximum possible distance of the ends of the molecules from the surface is reached. In spite of the variation in the film thickness during adsorption, the concentration of the polymer in it does not change. Calculations showed that during adsorption at the Θ-point the molecules are sorbed in conformations close to that of the random coil.

The experimental molecular sizes (that is, root-mean square thicknesses) are compared with the molecular sizes calculated according to the Di Marzio theory /162/. (The principle of the Di Marzio theory is that the conformation of the molecules on the surface can be described by random walks, while the surface is the reflecting or adsorbing barrier.) The results are shown in Table 16.

TABLE 16. Comparison between calculated and experimental values of molecular extension of adsorbed polystyrene molecules on a chrome-ferrotype surface

Mol. wt.	r, Å	$(\bar{z^2})^{1/2}_{ref}$	$(\bar{z^2})^{1/2}_{ads}$
76,000	170	110	170
207,000	290	190	270
537,000	420	300	440
1,370,000	530	440	630
1,900,000	740	580	810
3,300,000	830	760	1080

Note. $(z^2)^{1/2}$ is the root-mean square thickness calculated for the reflecting and adsorbing surface.

Grakin and his co-workers /60/ applied ellipsometry to the study of adsorption on finished and unfinished glass surfaces. The data obtained by ellipsometry for the structures of adsorbed polyester and polystyrene films are interesting /65/. It was found that the concentration of the polymer in the polyester film is 33—60 wt%, while the amount of polystyrene is only 10 wt%. The polymers were, however, not adsorbed from the Θ-solvent, which complicates processing of the data.

The conformation of polystyrene molecules adsorbed by chrome-ferrotype surfaces was studied and it was found that the film thickness depends on surface coverage /62/. This confirms the hypothesis on the transition from the flat to the extended conformation as adsorption proceeds. The equilibrium film thickness increases with increase in the concentration of the solution until the plateau is reached. This indicates the dependence of the conformation on the number of molecules competing for sites on the surface. The film thickness was approximately proportional to the square root of the molecular weight. This corresponds to the adsorption of macromolecules as random coils.

These data were compared with those on the adsorption of polystyrene on a metallic mercury surface /163/. It was found that the film thickness is independent of the molecular weight. The authors believe that the molecules at once acquire their final conformation on the mercury surface and this conformation remains constant during adsorption. The conformation is relatively flat, and does not pass into the extended, as the number of occupied sites increases. The authors believe that in this case the "holes" in the film are filled, while the concentration in the film increases to a limiting value.

It was also proposed that the pattern of interaction between the mercury and the polymer is such that the p value (fraction of attached segments) remains approximately constant and is independent of the molecular weight. The authors attempt to explain the dissimilarity in the behavior of the macromolecules during adsorption on the surface of solid metals and liquid mercury by the purity of the surface, adsorption of different gases on them, and by the difference in the contribution of the London dispersion forces to the surface energy of the mercury metal. Since the dispersion component of the free surface energy of mercury is much higher than for other metals, adsorption of the nonpolar polystyrene should apparently lead to a higher p and a smaller number of loops extending into the solution. The authors also consider that the homogeneity of the mercury surface is important, as well as the fact that the polymer interacts better under these conditions. The strong interaction, and thus the higher p value, lead to small changes in the film thickness during adsorption.

Stromberg et al. /63/ studied the conformation of adsorbed polystyrene macromolecules by the attenuated total reflection method. They obtained the data for adsorption from the Θ-solvent shown in Table 17. Their results agree with ellipsometric data.

It is worth noting that the film thicknesses for different surfaces with approximately the same free surface energies are very close. Silberberg /28/ used precision viscometry to estimate the film thickness, since he considered multilayer adsorption possible (formation of layers in which the macromolecules interact only with the segments of the other layer and not directly with the surface). According to these concepts the intermediate layer attached to the first may be considered as a new surface, similar to the surface of the concentrated gel of the same polymer /164/. In a sufficiently poor solvent this surface may be a good substrate for adsorption.

TABLE 17. Film thickness of polystyrene adsorbed at 34°C from cyclohexane according to ATR* data

λ, Å	r, Å	Ratio of concentrations in film and in solution
2690	230	39
2680	260	32
2670	240	34
2600	240	35

* [Attenuated total reflection.]

TABLE 18. Adsorption of polystyrene from toluene and cyclohexane at 34°C

Mol. wt.	$[\eta]$, cm³/g	$c \cdot 10^7$, g/ml	r, Å	Mol. wt.	$[\eta]$, cm³/g	$c \cdot 10^7$, g/ml	r, Å
Toluene						116.00	113
						245.00	136
$18.0 \cdot 10^5$	358	1.81	95			560.00	192
		3.67	118				
		5.13	161	Cyclohexane			
		6.94	192				
		8.52	251	$18.0 \cdot 10^5$	111	0.09	350
		10.22	273			1.77	920
$4.98 \cdot 10^5$	147	1.74	178			6.61	1060
		4.41	180			41.66	1080
		7.48	166			93.42	1040
		9.87	172	$4.98 \cdot 10^5$	60	1.57	730
		28.59	151			5.97	730
$2.60 \cdot 10^5$	85	75.70	85	$2.60 \cdot 10^5$	43	0.52	60
		91.30	96			5.00	140
						12.79	140
						125.85	140

Table 18 shows the data of Silberberg for anionic polystyrene solutions with a narrow molecular weight distribution at different concentrations. It can be seen that the film thickness is a function of the concentration of the solution, and that it increases with molecular weight. This may confirm multilayer adsorption. A very important paper on the structure of macromolecular films on solid surfaces is that of Killman and Welgand /61/. The authors studied the adsorption of poly (ethylene glycols) with molecular weights of 26,130 and 40,000 from aqueous solutions, and of poly (vinyl pyrrolidone) (molecular weight 38,000) from water and methanol. A chrome mirror was used as the adsorbent. The density and concentration of the polymer in the film were determined by ellipsometry, and from these data the amount of polymer sorbed by unit surface was calculated.

The same authors studied the dependence of the amount of adsorbed polymer and of the film thickness and concentration on time, temperature, concentration of solution, molecular weight of polymer, and nature of solvent (Figure 78) for solutions with an initial concentration of less than

10 mg/ml /61/. The conventional isotherms (with reference to the amount of adsorbed polymer) and the isotherms with reference to the film thickness and concentration in film c_2 were plotted (Figure 78). All the isotherms have a common characteristic shape. At concentrations of the solution above 1 mg/ml, the adsorption does not increase with increase in concentration, that is, saturation is reached. The film thickness also increases to a certain level only, depending on the solution parameters. The concentration in film c_2 decreases, and then changes symbatically with the concentration of the solution with a constant difference Δc (the concentration of the solution was determined from the refractive index). This excess concentration is considered to be the concentration of the adsorbed polymer.

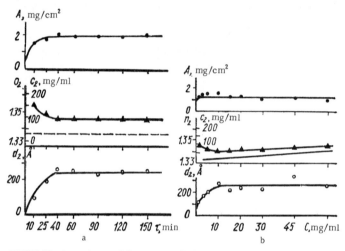

FIGURE 78. Dependence of the amount of adsorbed polymer, refractive index (concentration), and thickness of film on time (a) and concentration (b).

Killman and Welgand /61/ determined also the temperature dependence, and found that for both polymers the film thickness increases with increase in temperature, while the concentration in the film decreases. The data on the film thickness and concentration in the film were used to estimate the film structure and conformation of the adsorbed macromolecules. The authors state that the small film thickness and the high concentration in the film on the initial sections of the isotherms confirm the hypothesis that the molecules are spread out over the surface and are attached to it by a larger number of segments. At higher concentrations, the structure of the adsorbed film becomes rearranged. The newly adsorbed molecules will impair the existing bonds, so that the total number of bonds per adsorbed molecule will decrease. The molecule will straighten out on the surface and become displaced, the film thickness will increase, and the concentration in the film decrease.

At low concentrations, the competition for the sites on the surface is relatively weak, but as the concentration in the solution increases, the competition between the molecules will become stronger. Hence, the number of points of contact, that depends on the concentration, decreases. At a certain concentration, equilibrium between the competing molecules will be established, and the film thickness and concentration in it will no longer change to any extent. The adsorbed molecules will acquire a conformation that is very dependent on the interaction in the solution. The total concentration in the film will change symbatically with the concentration in the solution.

The authors note that when the film thickness is calculated from ellipsometric equations it is assumed that the structure of the film is homogeneous and that the concentration of the polymer in the film is constant. But these conditions do not hold in a real system, and averaged magnitudes are taken into account. However, the authors compared the film thickness with the chain sizes calculated from the intrinsic viscosity, and with the size of the maximally elongated chain. Table 19 gives the mean film thickness d_2, and the values of d_m corresponding to averaged Gaussian thicknesses, calculated by the method of Lipatov /165/.

TABLE 19. Correlation between film thickness and size of macromolecules

Parameter, Å	PEG-26,130 in water, 25°C	PEG-40,000 in water			PVP-38,000		
		25°C	35°C	45°C	in water, 25°C	in water, 45°C	in methanol, 45°C
d_2	258	254	247	320	650	725	635
d_m	157	154	150	194	394	440	385
$\sqrt{h^2}$	69	176	176	176	136	136	163
L_{max}	506	3300	3300	3300	863	863	863

The data of Table 19 show that the mean film thickness for PEG-40,000 corresponds to the size of the random coil, while for PEG-26,130 and PVP-38,000 it is between the size of the coil and the length of the maximally elongated chain. Actually, in the latter case the degree of polymerization is low, and therefore in solution the molecule does not have the configuration of a coil but of an ellipsoid. This may explain the difference between the sizes of the molecule and the film thickness. If the film thickness is equal to the distance between the ends of the chain for the molecular random coil of high molecular weight, adsorption may be in the form of coils, but the authors admit that adsorption in the form of loops is also possible, and accounts for the weak temperature dependence in the saturated film.

The problem of the conformation of adsorbed polymeric chains and its temperature dependence is very important, but there are very few experimental papers on the subject. Killman and Welgand /61/ processed the experimental data and concluded that the coil expands with temperature, so that the mobility of the molecular segments becomes higher. This indicates a potential spreading out of the coil with increase in temperature at low concentrations in solution, so that the film thickness at saturation corresponds to a growing end-to-end distance. The dependence of adsorption on molecular weight can be explained by considering that the molecules of higher molecular weight require a higher number of contact points with the surface for adsorption, and this corresponds to a smaller film thickness and higher film concentration. Data on the effect of the nature of the solvent determining the size of the polymeric molecules and their conformation can be explained similarly.

Perkel and Ullman /54/ studied the structure of the adsorbed film by using the dependence of polymer adsorption on molecular weight. The calculations were based on the models described above, and the difference in the dependence of the amount of the adsorbed polymer on molecular weight.

The following cases are examined:

1. All the polymeric segments lie on the adsorbent surface; the dependence of adsorption on molecular weight is described by $A_s \sim M^0$.

2. The polymeric molecules are attached to the surface by a single segment only: $A_s \sim M^1$. Adsorption is directly proportional to molecular weight.

3. The polymeric molecules are adsorbed as spheres, with a radius proportional to the radius of gyration of the molecules in the solution: $As \sim M/R^2$. Thus, if $R^2 \sim M^{1+2\alpha}, 0 < \alpha < 0, 1$, then $As \sim M^{-2\alpha}$. The value α is equal to zero for a poor solvent and 0.1 for a good one. When the coils interpenetrate, the density of the segments in the molecular coil increases. The increase is higher at higher molecular weights. Therefore As will change in proportion to higher powers of M.

4. The polymeric molecules are completely entangled and coiled on the surface. The density of the segments decreases with increase in the distance x from the surface:

$$\rho(x) = \rho_0 \exp(-Kx^2/M), \qquad (4.7)$$

where ρ_0 and K are constants independent of the molecular weight. In this case $As \sim M^{1/2}$.

5. The polymer is sorbed in two stages. In the first stage the molecules lie flat on the surface until complete coverage. Further adsorption of the polymer takes place only by attachment of several segments on the free sites of the surface; the molecules are coiled: $As = K_1 + K_2 M^b$. The constant term is determined by the molecules lying on the surface and attached to it by all segments, while M^b is determined by b (equal to 1 in case 2 or $1/2$ in case 4).

For the adsorption of polydimethylsiloxanes, Perkel and Ullman /54/ found that $As = KM^b$, where b lies between 0.2 and 0.45. These magnitudes are correlated to the various types of molecular distribution on the surface.

The same method for estimating the film structure was used in /111/ to study the adsorption of the ethylene-vinyl acetate copolymer on glass. It was found that α decreases with increase in molecular weight. This shows that the conformation of the adsorbed chain passes from the coiled to the extended structure with increase in molecular weight. This is explained by the increase in the number of polar segments in a single molecule that strongly interact with the surface.

A quantitative approach for estimating the structure of the adsorbed film was developed by Silberberg /166/. The author applied the following principles in his theory. If a flexible molecule is distributed over a flat surface, its shape differs from that in the bulk. This change is determined by the chemical nature of the structure of the chain. In an analysis of the structure of the adsorbed film, we may take into account the case when all the segments of the polymeric molecule are able to be adsorbed and all the points on the surface can adsorb, and the case when these conditions do not apply.

We shall consider a polymer molecule near the flat surface, and the conformation which it must attain so that at least one part of the segment is in contact with the surface. We divide the molecules into a sequence of segments which may or may not lie on the surface. In general, let m_i be sequences of i segments outside the surface, and $\overline{m}_j$ be sequences of j segments in the adsorbed state. This idea is simply the loop model considered above. Variables i and j may assume arbitrary values, but if the molecule has P segments, the general condition is that the following relationships hold:

$$P = \sum im_i + \sum j\overline{m}_j, \qquad (4.8)$$

$$\sum m_i = \sum \overline{m}_j. \qquad (4.9)$$

To solve the problem of the conformation of the adsorbed molecules, we must find functions $w\,(i)$ and $\overline{w}\,(j)$ which are the thermodynamic probabilities that the sequence of i segments forms a loop and the sequence of j segments is on the surface. These functions are found by calculating all potential configurations which the molecules may take up on the surface and in the loop. Thus, the authors of the theory found an expression for the partition functions of the whole system and the equilibrium conditions during adsorption.

Suppose that at equilibrium $i = P_B$ and $j = P_S$. The partition function then has its maximum value. From the general condition it follows that

$$m\,(P_S + P_B) = P, \qquad (4.10)$$

where m is the total run of the sequences of each kind. By introducing the equilibrium values P_B and P_S into the equation for the partition function, the author obtained a system of equations from which these values can be found. Silberberg assumed that the values of P_B and P_S convey complete information on the state of the polymer, since with them it is possible to calculate the fraction of the segments of the polymeric chain directly attached to the surface:

$$p = \frac{P_S}{P_B + P_S}. \qquad (4.11)$$

Then, the probabilities $W\,(P_B)$ and $\overline{W}\,(P_S)$ are found; they are calculated by assuming a certain type of lattice which describes the surface and the solution.

We shall consider this problem for a hexagonal lattice which is the simplest, but gives a complete idea of how this problem is solved in principle /166/. Suppose that the hexagonal lattice has a coordination number Z, that is, each point of the surface can be occupied by a segment surrounded by Z nearest neighbors. On the surface the coordination number decreases to S. Near the surface we may consider the lattice as an array of layers of sites similar to sites on the planar surface. Then we have $\frac{Z-S}{2}$ ways of stepping from one layer to the other, and S possibilities of staying on the surface. With these definitions, the problems of the number of arrangements on this lattice for a random walk when some of the segments set out from the surface and return to it can be handled in principle. The adsorption loop of P_B segments corresponds to a random walk of P_{B+1} steps from the surface, and vice versa. In this case

$$W(P_B) = \left(\frac{Z-S}{2}\right)^2 f^{P_B} (Z-1)^{P_B-1}, \qquad (4.12)$$

where f is a factor (per segment) expressing the effect of the restriction imposed on the loop by the surface in the hexagonal lattice.

Table 20 shows the calculation of the number of conformations for a different number of segments in the loop and number of layers considered in the lattice. If $Z = 12$ and $S = 6$, then by comparing equation (4.12) with the equations in Table 20, we can calculate f in each case (Table 20).

TABLE 20. Number of conformations of an adsorption loop on the Z—S lattice

Number of segments in loop P_B	Number of layers considered	Total number of conformations $W(P_B)$	f
1	1	$\left(\frac{Z-S}{2}\right)\left(\frac{Z-S}{2}-1\right)$	0.67
2	1	$\left(\frac{Z-S}{2}\right) S \left(\frac{Z-S}{2}\right)$	0.66
3	2	$\left(\frac{Z-S}{2}\right)\left(\frac{Z-S}{2}\right)\left(\frac{Z-S}{2}-1\right)\left(\frac{Z-S}{2}\right) +$ $+ \left(\frac{Z-S}{2}\right) S (S-1) \left(\frac{Z-S}{2}\right)$	0.72
4	2	$\left(\frac{Z-S}{2}\right)^2 S \left(\frac{Z-S}{2}\right)^2 + 2\left(\frac{Z-S}{2}\right)^2 \left(\frac{Z-S}{2} - 1\right) S \left(\frac{Z-S}{2}\right) + \left(\frac{Z-S}{2}\right) S (S-1)^2 \left(\frac{Z-S}{2}\right)$	0.72
5	3	$\left(\frac{Z-S}{2}\right)^3 \left(\frac{Z-S}{2}-1\right)\left(\frac{Z-S}{2}\right)^2 + \left(\frac{Z-S}{2}\right)^2 \left(\frac{Z-S}{2}-1\right)^3 \times$ $\times \left(\frac{Z-S}{2}\right) + \left(\frac{Z-S}{2}\right)^4 (2S^2 + S(S-1)) + \left(\frac{Z-S}{2}\right)^3 \times$ $\times \left(\frac{Z-S}{2} - 1\right)(2S(S-1)+S^2) +$ $+ \left(\frac{Z-S}{2}\right)^2 S (S-1)^3$	0.70

For probability $\overline{W}$ we can write

$$\overline{W}(P_S) = S(S-1)^{P_S-2} e^{xP_S}, \tag{4.13}$$

where x is the segment adsorption energy in κT units. Finally, an equation is obtained for P_S and P_B for various levels of adsorption energies

$$\ln \frac{P_S}{P_S-1} = \ln \frac{P_B}{P_B-1} + \ln \frac{Z-1}{S-1} + \ln f - x, \tag{4.14}$$

$$\ln \frac{1}{P_S-1} = \ln(P_B-1) + \ln \left\{ \left[\frac{1}{2}(Z-S) \right]^2 S/(Z-1)(S-1)^2 \right\}. \tag{4.15}$$

With these equations we can find P_S and P_B in the $Z-S$ lattice for different values of x, as shown in Figure 79. It can be seen that, for positive x, P_B is less than 5.5, that is, more than half of the segments are on the surface $(p > 0.5$ if $x > \frac{1}{2}\kappa T)$. Thus, the polymeric chain is, as it were, spread over the surface.

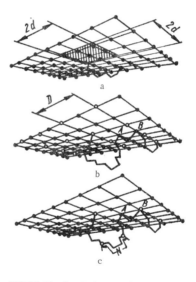

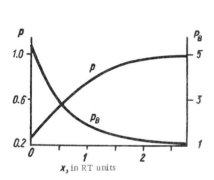

FIGURE 79. Dependence of the loop length P_B and of fraction p on the adsorption energy at $Z = 12$, $S = 6$, $f = 0.7$.

FIGURE 80. Restriction on adsorption loops in a cubic lattice.

Similar calculations can also be carried for a cubic lattice, where two choices of steps are considered: parallel and perpendicular to the surface plane.

In this case probabilities W_n and W_p are introduced, characterizing the number of configurations parallel and perpendicular to the surface. Depending on the restrictions imposed when these probabilities are found, the probability equations assume different forms. Some of these restrictions imposed on the chain configurations are shown in Figure 80.

Figure 80*a* shows the case when a return of the loop to the area is not allowed. The size of the surface area is determined by *nd*, where *d* is the unit of measurement of the distance in the lattice, equal to one half of the distance between the lattice points. Figure 80b shows the distribution of the molecules when not all the surface layer sites are adsorbing (those which are, are shown as open circles) and are distance *D* apart. Figure 80c shows the case when not all the segments are adsorbable (those which are, are shown as open circles). Therefore, in case 80b sites *A* and *B* can be occupied by segments without any changes in the energy of the system, while in case 80c point *A* is on the surface without energy change, while point *B* represents adsorption.

With these assumptions the authors calculated the fractions of the adsorbed segments *p* for different energies and the mean number P_B for various lattice types and parameters *d*, denoting the minimum distance separating the ends of the adsorbed loops (Figure 81). The thermodynamic probabilities $W(P_B)$ and $\overline{W}(P_S)$ were found for all these cases.

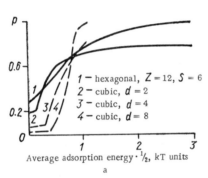
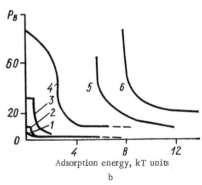

FIGURE 81. Dependence of the fraction of bound segments *p* for various lattice types (a), and of the mean number of segments P_B in the adsorption loop (b) on the adsorption energy.

Figure 81 shows the theoretical dependence of P_B on the adsorption energy for different cases. If all the sites on the adsorbent surface are adsorbing and all the segments are adsorbable, then the corresponding dependences for a hexagonal lattice ($Z = 12$, $S = 6$), a cubic lattice ($d = 2$), and a cubic lattice ($d = 4$) are described by curves 1–3. If only each tenth segment on the cubic lattice is adsorbable ($d = 2$), then the function is described by curve 4. For a cubic lattice ($d = 0$) on which only every tenth site is adsorbing while all the segments are adsorbable, the function is described by curve 5, while if only every tenth segment is adsorbable, the function is described by curve 6.

The authors arrived at the following conclusions. If the polymeric molecule is sufficiently large, contact with the surface will tend to split it into units. The size and structure of these are only determined by the nature and morphology of the surface, and are independent of the molecular weight. If all the sites on the surface are adsorbing, the segments are adsorbable, and the molecule is reasonably flexible, the loops will be short, and the molecules stay close to the surface, even if the adsorption energy is low. At about one κT energy change per interaction, adsorption is almost complete, with more than 70% of the segments in contact with the surface. Since the loops are short, and most of the segments are on the surface, the segment distribution in the vicinity of the surface has a maximum at the surface itself, so that it is possible to consider the adsorbed molecule as an essentially two-dimensional structure, which behaves rather like a macromolecule in a spread film.

The pattern depends, of course, on the structural and conformational restrictions imposed. The latter increase the loop size, film thickness, and energy required for adsorption. It is interesting to note that there is a region where P_B changes only slightly with x (Figure 81b). Since x is a measure of the intensity of interaction, it may be interpreted as, say, the inverse temperature. The data in Figure 81b (sharp change in the parameter) thus indicate the existence of a critical temperature at which desorption takes place. Hence a mixture of rather similar polymer species can be separated if we employ an adsorbent for which the x value for one polymer is slightly higher, and for the other polymer slightly lower, than the critical temperature. A separation on the basis of composition and structure rather than molecular weight would be achieved.

Silberberg /166/ was the first to show the important part played by adsorption processes and conformational restrictions in the interaction processes between polymeric molecules and enzymatic or catalytic surfaces, and also in polymerization processes on the interface, where the end of the growing chain is attached to the surface.

By analyzing the available published data, we can draw the following conclusions on the structure of the adsorbed film.

The film thickness and the conformation of the macromolecules in the film are determined by the number of points of contact with the surface. This number is larger at low concentrations of the solutions and low surface coverages. As the concentration of the solution increases and approaches the equilibrium value, the adsorbed film becomes rearranged, and the conformation of the adsorbed molecules changes correspondingly. Thus, we can consider that on saturation the film is formed by random coils, and is monomolecular in relation to these coils.

All the conclusions on the structure of the adsorbed film are qualitative. Changes in the conformation of the macromolecules themselves with changes in the concentration of the solution are not considered in even one paper on the conformation of the macromolecules in the adsorbed film. Conclusions can be drawn for infinitely dilute solutions only. But in real systems, the conformation of the macromolecules is concentration dependent, and therefore we must allow for changes in the conformation of the macromolecules during adsorption as the result of not only their interaction with their neighbors and changes in the number of points of attachment, but also intermolecular interaction in the solution. This aspect was first emphasized by us, and we shall now discuss it in detail.

Objections may be raised that in the elevated concentration range interpenetration and entanglement of these molecules begins as early as in the solution, and thus may not affect the conformation of the adsorbed molecules. This objection can, however, be discounted, since the concentration of the polymer in the adsorbed film is sufficiently high and the intermolecular action should be considerable.

We also noted that in some cases the amount of the molecule adsorbed by the surface is higher than that which could have been adsorbed if the molecules were in the form of random coils. A comparison of experimental and theoretical data is restricted, because all the existent theories were developed for low surface coverages. Many parameters (effective flexibility of the chains, coordination number, size of segments, number of contact points, and others) entering the theoretical equations are very difficult to determine, if at all, or they are too arbitrary for successful experimental verification.

It is worth noting that the general principles of the adsorption theory have been developed over a period when there was a lack of experimental data. Evidently, this situation in some way affects the general state of the theory of polymer solutions, which was thoroughly developed for dilute solutions only. We believe that several contradictions and confusions in the treatment of the adsorption and film structure can be eliminated if intermolecular action in solutions and on the surface is taken into account.

In some of our papers we have described the properties of monolayers of some oligomers and polymers. We can compare our data on the properties of monolayers with data on the structure of the adsorbed film because the model of monomolecular coverage by polymeric coils is adopted for these films also. In /32/ we studied the properties of the monolayers of low-molecular weight polyesters (oligomers) prepared from diethylene glycol and adipic acid as a function of the degree of polymerization. The data indicated a strong compressibility of the monolayer, and the existence of strong cohesion forces between the oligomeric molecules on the surface. It was also found that the packing of the monolayers is dependent on molecular weight. This indicates that both strong molecular interaction and conformational changes of the macromolecules may occur in the monolayer. It is, however, interesting that calculation of the surface area covered by the macromolecule gave an anomalously low value, while the thickness of the monolayer may exceed the length of the completely extended chain. Thus, we can conclude that in the given case the film thickness at high compression is determined by molecular aggregates formed, which are independent kinetic units.

It is obvious that molecular aggregates may also be formed in polymeric adsorbed films on solid surfaces, and these determine the structure of the film and its properties.

Chapter 5

STATISTICS AND THERMODYNAMICS
OF ADSORPTION

A very important problem in the study of physical adsorption of polymers is the establishment of a relationship between the thermodynamic functions characterizing the adsorption processes and the experimental functions obtained for various systems. This problem is much more complicated for high-molecular weight systems than for low-molecular weight ones, because a thermodynamic description of the flexible molecular chains in solutions, and of the many experimentally established features of polymer adsorption determined by the flexibility of the polymeric chains is complicated.

The specific features that complicate the thermodynamic description are the following: during the adsorption of polymeric molecules from a solution by a surface, the conformation of the molecules changes considerably, but the initial conformation depends largely on the nature of the solvent, the temperature, the concentration of the solution, and other factors. Therefore, an interpretation of the data is very dependent on the selection of the theoretical equation describing the system.

It is known /167/ that many completely differing models of adsorption processes lead to fairly similar theoretical isotherms. Therefore, the degree of coincidence between the theoretical and experimental isotherms is not a proof that the theoretical model or the parameters obtained from such a comparison are correct. The problem becomes even more complicated for polymeric systems, because in the analysis of the experimental data it is not always clear which surface of the adsorbent must be taken into account.

At present, the classical kinetic derivations of the equations for the isotherms are not being used even for the adsorption of gases on solid surfaces, because they do not allow for intermolecular reaction potentials in the system. Steele /167/ writes that the kinetic approach cannot apparently be used for the determination of the thermodynamic properties of a system determined by the intermolecular reaction potentials, and that the methods of statistical mechanics must be applied here.

Since statistical-mechanical methods have become very popular in the theory of polymeric solutions, the derivation of equations for the adsorption isotherms of polymers is based on the conformational statistics of the polymeric chains and the grand partition function for the given system in thermodynamic equilibrium.

In the general form for polymers, the problem is reduced to finding the sum of molecular states of the polymer in the adsorbed state and in solution. Since, as we know from statistical thermodynamics, the sum of the states is related to the thermodynamic functions, we can find an expression for the chemical potential of the polymeric molecule on the adsorbent surface and in solution, and then the equation of the adsorption isotherms.

However, before the existent theories on adsorption on which the equations of adsorption are based had been developed, numerous attempts were made to describe the adsorption of polymers by the classical equations derived for low-molecular weight systems.

THE LANGMUIR AND FREUNDLICH EQUATIONS
FOR THE ADSORPTION ISOTHERM

Because of the many differences between the properties of solutions of low-molecular weight substances and those of high-molecular weight substances, it would be rather surprising to find that the adsorption isotherms derived for low-molecular weight systems are applicable to high-molecular weight ones. Many attempts have, however, been made to apply the Langmuir equation for isotherms to polymeric systems. Langmuir's equation for an isotherm can be written

$$\Theta = \frac{G_p}{G_{pS}} = \frac{bc}{1 + bc},$$ \hfill (5.1)

where Θ is the surface coverage; G_p is the adsorbed amount at concentration c; G_{pS} is the same at saturation; b is a constant. The equation was derived by assuming that spherical molecules are in solution, and do not interact or change their shape during adsorption. Moreover, it was assumed that the system is thermodynamically reversible.

To evaluate the results obtained when studying adsorption, $1/G_p$ is plotted against $1/c$, or G_p/G_{pS} against $1/c$. Koral et al. /77/ showed that in the first case the points are distributed near the ordinates at high concentrations, while they are scattered at low concentrations.

If the adsorption results are expressed in the form of Langmuir's isotherm, they agree fairly well only at low concentrations /77, 98, 168, 169/. The results obtained at higher concentrations do not conform with Langmuir's isotherms, and the experimental data lie below the $1/G_p = f(1/c)$ line. Figure 82 shows data on the adsorption of poly(methyl methacrylate) on glass from toluene solutions. They show that Langmuir's equation is applicable in this case. Published data also confirm that at lower molecular weights of the polymer the adsorption isotherm can be described better by Langmuir's equation /39/.

Attempts were also made to apply the empirical Freundlich equation to describe adsorption

$$x = \beta c^\mu,$$ \hfill (5.2)

where x is the adsorption; c the equilibrium concentration of the solution; β and μ constants. Although Gyani /170/ found satisfactory agreement between the experimental data and Freundlich's equation for polysaccharides on activated charcoal, in the general case the agreement is poor, especially at low concentrations /171/.

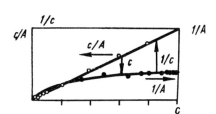

FIGURE 82. Applicability of Langmuir's isotherm to the adsorption of poly (methyl methacrylate) on glass at 30°C.

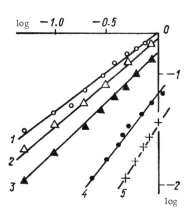

FIGURE 83. Isotherms of the adsorption of poly (methyl methacrylate) by fiber glass from solutions in acetone (1—3) and chloroform (4, 5) in coordinates of the Freundlich equation:

1) 60; 2, 4) 40; 3) 25; 5) 19°C.

However, we proved /38/ the applicability of Freundlich's adsorption isotherms over a fairly wide range of concentrations (Figure 83). We believe that the agreement between the experimental data and the equation can be explained by the fact that in the concentrational region studied we can no longer speak of the presence of isolated macromolecules. Their aggregates must be considered to be independent kinetic or structural units, passing upon the surface. In this case the adsorption mechanism is evidently no longer specific, because the configurational properties of the molecules do not exert such a strong influence as in dilute solutions.

ADSORPTION THEORY AND ADSORPTION ISOTHERM OF SIMHA – FRISCH – EIRICH

This theory is based on the following fundamental postulates /39, 172, 173/:

1. The solution of the polymer is assumed to be infinitely dilute.

2. It is assumed that the active centers are distributed regularly over the adsorbent surface, that their number N_s is proportional to the surface area of the adsorbent, and that the individual segments of the chain interact with the active centers.

3. The polymeric molecule is considered to be a more or less flexible random coil, characterized by a Gaussian distribution, consisting of t segments of length A.

4. The area of the active centers on the surface corresponds to the surface of the adsorbed segments and each active center can bind one segment only.

5. The mutual interaction of the polymeric molecules and the effect of the solvent on adsorption are characterized by two constants K_1 and K_2, while competitive adsorption of the solvent is not taken into account.

In random walk terminology, the surface of the adsorbent is considered to be a reflecting wall. When the adsorption isotherm is derived, the probability W of the end-to-end distance of the random chain in the solution and on the surface is found, assuming that the surface acts as a reflecting barrier. We suppose that the first chain segment is adsorbed at the beginning.

The probability $p\,(\tau)$ that of the segments of one chain the τ-th segment will be adsorbed again can be expressed by

$$p\,(\tau) = \frac{8\pi c^3 \alpha}{f^{3/2}\tau^{3/2}}\,(1 - e^{-\tau/4f}),\qquad (5.3)$$

where c is a normalizing factor $(c = 2\pi^{1/2})$; f is a measure of the chain flexibility; α is the probability that the segment lying on the surface is attached to the surface by adsorption forces.

The average probability p for all t segments is expressed by

$$p = \langle p\,(\tau)\rangle_t = \frac{1}{t}\sum_{\tau=1}^{t} p\,(\tau) = \frac{1}{t}\int_{1}^{t} p\,(\tau)\,d\tau,\qquad (5.4)$$

and since at high values of t, $(p\,(\tau) - p)$ is small, $p\,(\tau)$ can be replaced by p.

From these assumptions we can calculate the probability $u\,(\nu,\,t)$ that a molecule consisting of t segments will be attached to the surface by segments

$$u\,(\nu,\,t) = \frac{t!}{(t-\nu)!\,\nu!}\,p^{\nu}\,(1-p)^{t-\nu}.\qquad (5.5)$$

The number of adsorbed segments ν is determined by

$$\langle \nu \rangle = \sum_{\nu=1}^{t} \nu u\,(\nu t) = pt.\qquad (5.6)$$

If N_A molecules are adsorbed, a total of N_ν segments are adsorbed:

$$N_\nu = N_A u\,(\nu,\,t) = N_A \frac{t!}{(t-\nu)!\,\nu!}\,p^{\nu}\,(1-p)^{t-\nu}.\qquad (5.7)$$

Moreover, it is assumed that the adsorbed molecule is formed by subchains consisting of $1,2,\,\ldots,\,\lambda$ adjacent segments connected by bridges, which are not adsorbed and which stick out of the adsorbed surface. The probability of adsorption of the λ-mer (train of λ adjacent adsorbed segments) for long chains will be $p^{\lambda-1}(1-p) \approx p^{\lambda-1}$, since $p \ll 1$. In a similar manner we can assume that all the adsorbed chains are attached to the surface by $\langle v \rangle$ segments. In this case the number of λ-mers will be expressed by

$$M_\lambda = N_A \langle v \rangle (1-p) \, p^{\lambda-1}(1-p) = N_A p^\lambda (1-p)^2, \qquad (5.8)$$

since $\langle v \rangle = pt$.

We should note that p is the fraction of attached segments and can be found experimentally, which enables us to assume the conformation of the adsorbed chains. The size distribution M_ε of the bridges and the number-average chain length $\langle \varepsilon \rangle$ of a bridge are given by

$$M_\varepsilon = N_A p^2 t (1-p)^\varepsilon \qquad (5.9)$$

and

$$\langle \varepsilon \rangle = t/\langle v \rangle - 1 = (1-p)/p. \qquad (5.10)$$

These concepts were necessary for calculating the thermodynamic properties of the adsorbed film. The mole fractions n_0 of the empty sites and those occupied by λ-mers (n_λ) are determined by

$$n_\lambda = M_\lambda/N_S = N_A p^\lambda (1-p)^2 \, t/N_S , \qquad (5.11)$$
$$n_0 = 1 - \sum_\lambda M_\lambda/N_S = 1 - N_A pt (1-p)/N_S. \qquad (5.12)$$

Thus, we can readily obtain the expressions for the volume fractions

$$\varphi_0 = 1 - \sum_\lambda M_\lambda \lambda/N_S = 1 - N_A pt/N_S,$$
$$\varphi_\lambda = \lambda M_\lambda/N_S = N_A \lambda p^\lambda t (1-p)^2/N_S. \qquad (5.13)$$

Then, we must find the partition function of the whole system, which for N polymeric molecules, N_A of which are adsorbed, can be expressed by

$$Z_f = Z_{\text{mi}}^\lambda Z_{\text{mi}}^\varepsilon e^{N_A \langle v \rangle x/kT} \left[\prod_{i=1}^{t} j_i (T) \right]^N [j_S (T)]^{\langle v \rangle N_A} (j_t)^{N-N_0} Z_{\text{mi}}(N - N_{A\varepsilon}), (5.14)$$

where Z_{mi}^λ describes the mixing of the λ-mers with the empty sites; $Z_{\text{mi}}^\varepsilon$ the mixing of the ε-bridges with the solution; x is the adsorption energy of the segment; $j_i (T)$, $j_t (T)$ and $j_S (T)$ are the partition functions of the internal degrees of freedom of the segments, the translational degrees of freedom in solution, and the vibrations around adsorption sites, respectively. The term $Z_{\text{mi}}(N - N_{A,\varepsilon})$ characterizes the mixing in the adsorbed film of $N - N_A$ dissolved molecules, M_ε bridges, and M_L solvent molecules.

The partition function of the system before adsorption is determined by

$$Z = Z_{\text{mi}}(N) \left[\prod_{i=1}^{t} j_i (T) \right]^N (j_t)^N, \qquad (5.15)$$

from which we obtain the change in the free energy of the system during adsorption

$$\Delta F = F_f - F = kT \ln (Z_f/Z). \qquad (5.16)$$

Explicit expressions for the mixing terms are obtained from the Flory – Huggins theory of polymeric solutions. If we introduce

$$\Theta = \sum_\lambda \lambda n_\lambda = N_A \langle v \rangle / N_S \qquad (5.17)$$

(Θ is the surface coverage), then from (5.14) we can write the free energy of the adsorption system

$$-kT \ln Z_f = kTN_S \{(1-\Theta) \ln (1-\Theta) + \\
+ \Theta \ln \Theta + p\Theta \ln [p (1-\Theta)/\Theta] + K_1\Theta^2\} + \\
+ kTL [\ln (L/(L + (t - \langle v \rangle N_A)) + \\
+ K_2 (t - \langle v \rangle)N_A/(L + (t - \langle v \rangle) N_A)] - \\
- N_A \langle v \rangle x - kTN_A \sum_i \ln j_t (T) - kT \langle v \rangle N_A \ln j_s (T). \qquad (5.18)$$

From this equation we can readily obtain the partial molar free energy of the adsorbed polymer by differentiating with respect to N_A:

$$\mu_f/kT = 1/kT\partial F_f/\partial N_A, \qquad (5.19)$$

By comparing the chemical potentials of the polymer in the adsorbed film and in the solution we can obtain the equation of the adsorption isotherm

$$\{[\Theta e^{2K_1\Theta}/(1-\Theta)] [p (1-\Theta)/\Theta e^{-1/(1-\Theta)}]^p\}^{\langle v \rangle} = K_{(T,t,x,K_2)}c, \qquad (5.20)$$

$$K = (M_L/M_S d) e^{\langle v \rangle/kT [x-kT(1-K_2)]} \frac{h^3}{(2\pi mkT)^{3/2}} j_S^{\langle v \rangle} \bar{V}^{-1}, \qquad (5.21)$$

where $\bar{V}$ may be regarded as the "free volume" per molecule of solid: M_L, M_S are the molecular weights of the solvent and the segment, respectively; d is the density of the solvent; h is Planck's constant.

Equation (5.20) holds for the case when each macromolecule is, on the average, attached by $\langle v \rangle$ segments. The equation does not explicitly contain parameter f, characterizing the flexibility of the chain; it can, however, be proved that $\langle v \rangle = pt \sim f^{-1/2}t^{1/2}$.

In a simpler form, if we take $\lambda = 1$, the equation of the isotherm is transformed into

$$[\Theta e^{2K_1\Theta}/(1-\Theta)]^{\langle v \rangle} = Kc. \qquad (5.22)$$

At $\langle v \rangle = 1$, (5.20) is transformed into Langmuir's equation, since $K_1\Theta \ll 1$.

Using equation (5.17), the free energy of adsorption can be expressed by

$$\Delta F/kTN_S = (1-\Theta) \ln (1-\Theta) + \Theta \ln \Theta + p\Theta \ln [p (1-\Theta)/\Theta] + \\
+ (\Theta/\langle v \rangle) \ln j_t - \Theta \ln j_s - \Theta (x/kT + K_2) + K_1\Theta^2, \qquad (5.23)$$

and the enthalpy by

$$\Delta H/kT = -N_A\langle v\rangle (x/kT + K_1') + (K_2'N_A^2/N_S)(\langle v\rangle)^2, \qquad (5.24)$$

where K_1' and K_2' are the interaction constants in the adsorbed film and in the solution.

We shall now consider the character of the Simha – Frisch – Eirich isotherm. At $\langle v\rangle > 1$, this isotherm is a curve with rapid initial ascent and smooth attainment of equilibrium. The saturation line lies below the line given by the Langmuir equation, corresponding to $\langle v\rangle = 1$ and $\lambda = 1$.

Very interesting conclusions can be derived from (5.20). In poor solvents the flexibility of the chain f is low (we should note that we always refer to kinetic and not thermodynamic flexibility), that is, $\langle v\rangle$ will be small. The interaction constant K_2 is also lower than in a good solvent. Both effects increase constant K, and determine a better adsorption from a poor solvent. The temperature dependence of adsorption is determined by the bond energy x of the adsorbed segment, the interaction constant K_1, the number of different degrees of freedom of the adsorbed molecule, the flexibility of the chain f, and the magnitude $\langle v\rangle$ related to it. All these factors may lead to both an increase and a decrease in adsorption with increase in temperature.

When the equation for the isotherm was derived, competitive adsorption of the solvent was not taken into account. Simha and Frisch /174/ showed that if we neglect the solvent-polymer interaction, the adsorption of the solvent does not affect the character of the isotherm, except that the saturation state is reached at higher concentrations in the solution. The authors also allowed for the interaction of the polymeric molecules, and introduced the concept of the reflecting wall E, as the result of which the adsorbed segments form obstacles for further adsorption. The height of the wall is determined by the number of loops that restrict access to the adsorption centers, and is a function of the surface coverage Θ. Because of the above interaction, $\langle v\rangle$ can be expressed as

$$\langle v\rangle = pt = [2\alpha t^{1/2}/(\pi f)^{1/2}][1-\Theta][1-([E]/kT)(\langle\varepsilon\rangle\Theta/l)], \qquad (5.25)$$

where $l \sim \langle\varepsilon\rangle^{1/2} \sim t^{1/4}$ is the width of the wall. The interaction increases with increase in molecular weight and chain flexibility, and the barrier increases with increase in temperature and greater solvent power.

MULTILAYER ADSORPTION

Frisch and Simha discussed the problem of multilayer adsorption, and postulated the following:

1. The adsorption center can bind s layers from the segments.
2. Horizontal interactions are negligible.
3. Participation of the solvent in adsorption can be neglected.
4. The polymers are adsorbed via $\langle v\rangle$ separate loops, independently of one another. Magnitude $\langle v\rangle$ is equal to the mean number of bonds of the individual molecule in all layers.

For the last case, the equation of the isotherm will be

$$N_A = t k v_m c^{1/\langle v \rangle} / \{ \langle v \rangle \, (c_S^{1/\langle v \rangle} - c^{1/\langle v \rangle}) \, [1 + (k-1) \, (c/c_S)^{1/\langle v \rangle}] \}, \qquad (5.26)$$

where N_A is the number of adsorbed chains; c_S is the saturation concentration; K is a constant; $v_m = S\sigma_0$ is the number of adsorbed segments at monomolecular surface coverage; σ_0 is the area occupied by the segment. The shape of this isotherm is similar to that of the BET isotherm for gas adsorption.

Changes in the state of the random coil during adsorption were taken into account in the later theory. Frisch /175/ introduced the idea of the adsorption potential, that acts normally to the surface. It is the probability that a polymeric segment will be present in a certain volume of the adsorption space. From the state in which all the molecules are attached to the adsorbent by a single segment, the probability of further segment adsorption is calculated. In this case the adsorption isotherm is

$$\Theta/(1-\Theta) = K c^{1/\langle v \rangle} s = K c^{1/\langle v \rangle [1 + \lambda_0]}, \qquad (5.27)$$

where $\langle v \rangle_s = \langle v \rangle \, [1 + \lambda_0]$; $\lambda_0 = \varepsilon_0 / kT$; ε_0 is the adsorption potential for weak adsorption interaction. If the interaction is strong, all the segments should be attached to the surface, and even at a very small concentration of the solution saturation should be achieved. However, such a case was not experimentally found.

Frisch and Simha /176, 177/ studied the configurational possibilities of the macromolecules on the surface. They started from the statistical theory of solvents of Guggenheim and Miller, and calculated the probability for an arbitrary conformation, determined the partition function, and compared the chemical potentials. They found the equation

$$\frac{\Theta \, [1 - (2/S) \, (1 - 1/t) \, \Theta]^{\langle v \rangle - 1}}{\langle v \rangle \, (1 - \Theta)^{\langle v \rangle}} = K_k c,$$

$$K_k = \frac{M_L}{M_S d} \, e^{\frac{\langle v \rangle x}{kT}} \, \frac{h^3 j_S^{\langle v \rangle} T \rho_S}{(2\pi m k T)^{3/2} \tilde{V}} \left[1 - \frac{2(t-1)}{tZ} \right]^{q_Z Z/2}, \qquad (5.28)$$

where x is the energy of adsorption of the segment; Z is the coordination number of the three-dimensional solvent lattice; S is the coordination number on the surface· K_k is a constant; M_L and M_S are the molecular weights of solvent and segment; d is the density of the solvent; j_S is the partition function of the internal degrees of freedom of the adsorbed segment; $q_Z Z$ is the number of possible neighboring segments of the same molecule; ρ_S is a configuration factor. Since $S, Z \gg 1$, we have

$$\left[\frac{1 - \Theta + \langle v \rangle \, \Theta}{\Theta \, (1 - \Theta)} \right] \frac{d\Theta}{dt} = \frac{d \ln \langle v \rangle}{dt} + \ln (1 - \Theta) \frac{d \langle v \rangle}{dt} + \frac{d \ln K_k \langle v \rangle}{dt}, \qquad (5.29)$$

that is, the surface coverage increases with increase in molecular weight. The temperature dependence of the adsorption is described by

$$\frac{1 + (\langle v \rangle - 1)\,\Theta}{\Theta\,(1 - \Theta)} \cdot \frac{d\Theta}{\partial T} = \ln(1 - \Theta)\frac{d\,\langle v \rangle}{dT} + \frac{d\ln\,(K_k\,\langle v \rangle)}{dT}, \qquad (5.30)$$

which indicates that surface coverage may both increase and decrease with temperature.

These isotherms were derived under the condition of thermodynamic equilibrium, but they can also be derived kinetically from adsorption and desorption rates /98/. Suppose the surface contains N_S adsorption centers able to bind one chain segment each. The polymer contains N molecules from t segments (out of which v segments are attached to the surface at $v < t$) and N_0 solvent molecules. The surface coverage by the segments Θ and solvent molecules Θ_0 is determined by

$$\Theta = vN'/N_S, \qquad (5.31)$$

$$\Theta_0 = N_0'/N_S, \qquad (5.32)$$

where N' is the number of adsorbed macromolecules; N_0' is the number of adsorbed solvent molecules. The number of molecules remaining in solution of concentration c (g/ml) will be $(N - N')$, and the number of solvent molecules at a concentration of solvent in solution c_0 (g/ml) will be $(N_0 - N_0')$.

We shall examine the following equilibria:

$$\text{Macromolecules in solution} + v \text{ free sites} \underset{K_2}{\overset{K_1}{\rightleftarrows}} N' \text{ adsorbed macromolecules, held at } v \text{ segments.}$$

$$\text{Molecules of solution} + 1 \text{ free site} \underset{K_2^0}{\overset{K_1^0}{\rightleftarrows}} N_1^0 \text{ adsorbed solvent molecules.}$$

Then, the number-concentration of "free sites" will be $1 - \Theta - \Theta_0$).

By applying the law of mass action, we obtain:
rate of polymer deposition,

$$r_1 = K_1 c\,(1 - \Theta - \Theta_0)^v; \qquad (5.33)$$

rate of polymer re-solution,

$$r_2 = K_2 \Theta/v; \qquad (5.34)$$

rate of solvent deposition,

$$r_1^0 = K_1^0 c_0\,(1 - \Theta - \Theta_0); \qquad (5.35)$$

rate of solvent re-solution,

$$r_2^0 = K_2^0 \Theta_0. \qquad (5.36)$$

At equilibrium, by comparing r_1 and r_2 and dividing the two sides of the equation by $(1 - \Theta)^v$, we obtain

$$\Theta/v\,(1 - \Theta)^v = Kvc\,[1 - \Theta_0/(1 - \Theta)]^v, \qquad (5.37)$$

where $K = K_1/K_2$,

$$\Theta_0/(1 - \Theta) = K_0 c_0/(1 + K_0 c_0) = \beta_0 \leqslant 1, \tag{5.38}$$

where $K_0 = K_1^0/K_2^0$. By dividing the equations by $(1 - \Theta)$, and carrying out some conversions, we obtain

$$\Theta/v\,(1 - \Theta)^v = (1 - \beta_0^v)\,K_v c = Kc. \tag{5.39}$$

At $K_0 \to 0$, (5.38) is reduced to the Simha — Frisch equation (5.20), and at $K_v \to 0$, (5.39) reduces to the expected Langmuir isotherm.

The kinetic derivation of the isotherm was somewhat modified by Fontana /178/, who introduced the blocking factor f into the expression for $\Theta = \frac{f v N^1}{N_S}$, which takes into account that each bound segment blocks f adsorption centers (including the point of attachment). In this case factor f enters into the right-hand side of (5.38).

ADSORPTION ISOTHERM OF
POLYDISPERSE POLYMERS

The theory of Simha — Frisch — Eirich did not take into account the poly-dispersity of the polymer in adsorption. The equilibrium adsorption of polydisperse polymers was theoretically considered by Gilliland and Gutoff /107/. In this case the derivation of the adsorption isotherms consists also in finding and comparing the partial free energies of the adsorbed and dissolved molecules in equilibrium. The following was postulated when the equation was derived.

1. The solvent molecules, chain segments, and the adsorption centers are described by one lattice model; the lattice model of Flory is adopted.

2. The concentration of the solution is so low that the polymer — polymer interaction can be neglected.

3. The solution is at the Θ -temperature.

4. The density of the polymer on the adsorbent surface is considered to be equal to that in the adsorbed film.

The free energy change when a pure amorphous polymer is mixed with a pure solvent will be

$$\frac{\Delta F_{mi}}{kT} = N_0 \ln v_0 + \sum_t u_t \ln v_t + K_2 N_0 v_2, \tag{5.40}$$

where N_0 is the number of solvent molecules; v_0 is the volume fraction of the solvent; v_t is the volume fraction of the polymer of dimension t; u_t is the number of such molecules; v_2 is the total volume fraction of the polymer; t is the number of segments of the polymeric chain with a segment of volume equivalent to the solvent molecule: K_2 is a measure of close interaction of the segment with the solvent.

By differentiating with respect to u_t, we can find the partial free energy of the solution

$$\frac{\partial \Delta F_{mi}}{kT \partial u_t} = \ln v_t - t + 1 + v_2 t - v_t + K_2 t (1 - v_2)^2. \qquad (5.41)$$

The partial free energy of the adsorbed polymer in the adsorbed film is found for two cases.

Two-dimensional case. All the segments of the adsorbed molecule lie on the surface. If A_t molecules of size t are adsorbed, and the heat of adsorption is equal to H (ergs/segment), then the change in free energy as the result of the thermal effect of the transition from the pure amorphous polymer to the adsorbed one will be

$$\frac{\Delta F^{(h)}}{kT} = - \sum_t A_t t H / kT \qquad (5.42)$$

and the partial free energy will be

$$\frac{\partial \Delta F^{(h)}}{kT \partial A_t} = - t H / kT. \qquad (5.43)$$

For mixing the adsorbed molecules with the surface

$$\frac{\Delta F^{(a)}}{kT} = N_S (1 - \Phi) \ln (1 - \Phi) + \sum_t A_t \ln \Phi_t + N_S K_1 \Phi (1 - \Phi), \qquad (5.44)$$

where N_S is the number of adsorption sites sufficient for the adsorption of one segment; K_1 is a parameter expressing the energy of near interaction, divided by kT for the adsorbed segments; Φ is the fraction of sites occupied by the polymer. Then

$$\Phi_t = A_t t / N_S, \qquad (5.45)$$
$$d\Phi_t = (t / N_S) \, dA_t, \qquad (5.46)$$
$$\Phi = \sum_t \Phi(t), \qquad (5.47)$$
$$\partial \Phi / \partial \Phi_t = 1, \qquad (5.48)$$
$$\frac{\partial \Delta F}{\partial A_t} = \frac{t}{N_S} \cdot \frac{\partial F}{\partial \theta}. \qquad (5.49)$$

Hence

$$\frac{\partial \Delta F^{(a)}}{kT \partial A_t} = t [K_1 - 2 K_1 \Phi - 1 - \ln (1 - \Phi)] + \ln (A_t t / N_S) + 1. \qquad (5.50)$$

During adsorption, the rotational, vibrational, and translational motions of the molecule will change. The change in free energy because of change in these motions during adsorption can be expressed by

$$\frac{\Delta F^{(i)}}{kT} = -\sum_t A_t t \ln j_a + \sum_t A_t t \ln j_u, \qquad (5.51)$$

$$\frac{\partial \Delta F}{kT \partial A_t} = -t \ln \frac{j_a}{j_u}, \qquad (5.52)$$

where j_a and j_u are the numbers of degrees of freedom of the polymer in the adsorbed film and in the solution, respectively. The equilibrium conditions can be found by comparing the partial free energies of the dissolved and adsorbed polymer. After some transformations, we obtain

$$\frac{A_t}{u_t} = \frac{N_S}{N_0}\left[\frac{j_a}{j_u}(1-\Phi)\exp\{H/kT + K_2 - K_1(1-2\Phi)\}\right]^t. \qquad (5.53)$$

Three-dimensional case. It is assumed that some of the segments of the adsorbed molecules are not adsorbed and extend into the solution, while others are adsorbed on the adsorbent surface. When the isotherm is derived, the mixing of the adsorbed segments with the adsorption points and of the nonadsorbed segments with the solution is taken into account.

Gilliland and Gutoff consider the lattice model, including the surface points, extending into the solvent to such a distance that the fraction of occupied sites in the lattice is equal to the fraction of occupied sites on the surface. The final equation is

$$\frac{A_t}{u_t} = \frac{N_S}{N_0 p}\{(1-\Phi)(j_a/j_u)^p \exp[pH/kT - K_2 - K_1'(1-2\Phi)]\}^t, \qquad (5.54)$$

where p is the probability of the adsorption of any segment of the adsorbed molecule on the surface; this probability is inversely proportional to the molecular weight

$$p = q/t^{1/2}.$$

Thus, (5.54) is transformed into

$$\frac{A_t}{u_t} = \frac{N_S t^{1/2}}{N_0 q}[(1-\Phi)e^{K_2 - K_1'(1-2\Phi)}]^t ((j_a/j_u)e^{H/kT})^{qt^{1/2}}. \qquad (5.55)$$

The good qualitative correspondence of the rough general model and of the more accurate two-dimensional model shows that the models are well correlated.

Equations (5.53), (5.54) or (5.55) show that under certain conditions the adsorption proceeds so that the polymer can be fractionated, and the molecular weight distribution curves can be determined directly by a simple equilibrium experiment.

If the term raised to the t-th power is greater than unity, then the high-molecular weight fractions are preferentially adsorbed. With this method, Gilliland and Gutoff /107/ determined the molecular-weight distribution of polyisobutylene and butyl rubber dissolved in benzene, using channel black as the adsorbent.

The isotherm suggested can be generalized for the adsorption of a homogeneous polymer

$$\Phi = A_t pt/N_S,$$ (5.56)

where $p = 1$ for the two-dimensional case, and $q/t^{1/t}$ in the general case. The concentration of the solution (g/100 ml) is

$$c = \frac{u_t t M_S \rho_0 (100)}{N_0 M_0}$$ (5.57)

(M_S, M_0 are the molecular weights of the segment and solvent; ρ is the density). Equation (5.33) reduces to

$$\Phi \left[\frac{e^{K_1(1-2\Phi)}}{1-\Phi} \right]^t = c \left\{ \frac{M_0}{100 M_S \rho_0} \left[\frac{j_a}{j_u} e^{(H/kT+K_s)} \right]^t \right\}.$$ (5.58)

In the general case (5.54) is transformed into

$$\Phi \left| \frac{e^{K_1(1-2\Phi)}}{1-\Phi} \right|^t = c \left\{ \frac{M_0}{100 M_S \rho_0} \left[\frac{j_a}{j_u} e^{(pH/kT+K_s)} \right]^t \right\}.$$ (5.59)

From an analysis of the isotherms (5.59), we can conclude the following. An increase in the heat of adsorption should be accompanied by preferential adsorption of the higher molecular weight fractions. A temperature increase leads to a decrease in the amount of adsorbed polymer.

However, this equation applies at the Θ-temperature only, and does not make it possible to predict the effect of the solvent on adsorption. The Gilliland-Gutoff equation was confirmed by taking the adsorption of rubbers on carbon black near the Θ-temperature as an example. The theoretical assumptions were qualitatively confirmed. An analysis of the Simha $-$ Frisch $-$ Eirich and the Gilliland $-$ Gutoff isotherms shows that these are fairly close to one another (see equations (5.32) and (5.59)).

SILBERBERG'S THEORY

Silberberg /166, 179—181/ contributed most to the theory of polymer adsorption. The author examined the problem from all points of view, and laid the foundations for further advance in theoretical concepts.

The theoretical description of adsorption is based on Silberberg's concepts on the structure of the adsorbed film, which was partially explained in Chapter 4. The author derived the equation of the adsorption isotherms by assuming that a train of bound segments are formed on the surface, which pass into loops extending into the solution. The author notes that the SFE theory predicts loops that are too long and films that are too thick, and cannot explain the large values of parameter p or the molecular weight dependence of the adsorption. Silberberg /167/ considered the behavior of an isolated molecule on the surface, and showed that its shape

is independent of the length of the molecule, and is determined only by
the energy of adsorption and steric factors, correlated to the lattice type.
To obtain the isotherm, Silberberg introduced a model that differs somewhat
from the previous ones and accounts more accurately for the experimental
facts.

Figure 84 shows a diagram of the adsorption of a polymer on a surface.
If there are m trains of segments bound to the surface in one molecule,
there must be $(m-1)$ trains pending in solutions. Let P_S segments per
train be on the surface and P_B segments per train be in the bulk of the
solution. It is, however, convenient to consider four types of segments in
the polymeric molecule. We shall assume that there are t segments which
are adsorbed, and stand at the head of the train (black squares), and
segments that are adsorbed, but do not stand at the head of the train (black
circles). Then there will be $(m + 1)$ molecules at the head of the loop
(open squares), and u segments inside the loop (open circles). If there is
a total of P segments per molecule, we have

$$P = 2m + 1 + u + t. \tag{5.60}$$

Since P, m, u, and t are large, the 1 can be neglected. With the new definitions (Figure 84)

$$P_S = \frac{m+t}{m}, \tag{5.61}$$

$$P_B = \frac{m+1+u}{m+t}, \tag{5.62}$$

whence

$$p = \frac{m+t}{P} = \frac{mP_S}{P} \cong \frac{P_S}{P_S + P_B}, \tag{5.63}$$

where p is the fraction of attached segments.

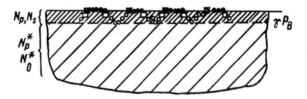

FIGURE 84. Schematic diagram of a polymer adsorbed on a
surface.

We shall now assume that the lattice has N_S adsorption sites and that N_p
polymer molecules are adsorbed by them. We shall denote the number of
unbound molecules by N_p^*. Then, the total number of molecules of the
polymer is obviously

$$N_{pT} = N_p + N_p^*. \tag{5.64}$$

Since the overall concentration of the solution is much lower than the concentration on the interface, it is best to consider the part of the volume lattice into which the loops penetrate as a separate solution of higher concentration. We shall denote the number of layers of lattice sites near the surface by γP_B (it is best to take γ as equal to $1/2$, in accordance with the figure). Then,

$$N_B = \gamma P_B N_S \tag{5.65}$$

is the number of lattice sites in the surface adjacent to the bulk phase (N_S is the number of sites in the surface layer). Thus, the volume fraction of the polymer segments in the bulk phase Φ is defined as

$$\Phi = \frac{(1-p) P N_p}{N_B}. \tag{5.66}$$

It is assumed that some of the segments belonging to N_p^* nonadsorbed molecules may occupy some of the N_B sites. If Φ^* is the volume fraction of the segments in the equilibrium solution, then $\Phi^* \ll \Phi$ is the limiting condition for the model. If N_0^* is the number of solvent molecules (situated below the dotted line in Figure 84) we have

$$\Phi^* = \frac{P N_p^*}{P N_p + N_0^*}, \tag{5.67}$$

and thus the total number of solvent molecules in the system is

$$N_{oT} = N_0^* + (N_S + N_B - P N_p). \tag{5.68}$$

The fraction Θ of the N_S surface sites occupied by polymeric segments is given by

$$\Theta = \frac{p P N_p}{N_S} = \frac{(m + f) N_p}{N_S} = \gamma P_S \Phi \tag{5.69}$$

and should be clearly distinguished from the total number Γ of adsorbed segments on the surface

$$\Gamma = \frac{P N_p}{N_S} = \frac{\Theta}{p}. \tag{5.70}$$

Thus, irrespective of the introduction of many different parameters, all of them may be reduced into terms of seven fundamental variables: N_S, N_p, N_p^*, N_0^*, m, u, t.

However, for any given system, P, N_{P_T}, N_{oT} are constant, and (5.60), (5.63) and (5.64) will reduce to a system with four independent variables. When we consider equilibrium in the system, N_S can be taken as constant, and we have three determining equations. Two of them are similar to the equations for P_B and P_S, and the third is the equation for the adsorption isotherm. If we allow for the above variables, we can write the partition functions of the whole system

$$Q_T = Q^* Q_S, \tag{5.71}$$
$$Q^* = \sum \sum Q^*(N_p^*, \; N_0^*), \tag{5.72}$$
$$Q_S = \sum \sum \sum \sum Q_S(N_P, \; N_S, \; m, \; t, \; u), \tag{5.73}$$

where Q^* and Q_S are the partition functions for the bulk and surface phases, and Q^* and Q_S are the general terms in these functions.

When we calculate these magnitudes, we consider the adsorption energy of polymeric segment x. When we calculate Q_T, we take into account not only the configurational changes, but also the depths of the interaction potentials of the three types: polymer – polymer (V_{pp}), polymer – solvent (V_{op}), and solvent – solvent (V_{oo}).

$$x_0 = \frac{z}{2} [V_{op} - V_{oo}] \frac{1}{kT}. \tag{5.74}$$

The change in the internal energy of the system during adsorption is determined by the change in the solvation energy during adsorption x and the parameter x_0.

We can calculate the number of conformations in the bulk and in the surface phase according to the scheme shown in Figure 84, then the partition functions, and find the third equation. The final result based on this model has the form

$$G^* = P\Delta + H, \tag{5.75}$$

where

$$G^* = \ln \left[\frac{(1 - \Phi^*) + q\Phi^*}{\Phi^*} \right] + P \ln \left[\frac{(1 - \Phi^*)}{(1 - \Phi^*) + q\Phi^*} \right]; \tag{5.76}$$

Δ and H are complex functions of parameters Z, S, P_B, P_S, p, m, etc., while q is a magnitude characterizing the contacts of the segments in the solution phase.

The adsorption equation in the complete form is complex and contains some magnitudes which cannot be experimentally determined without correlation with the model used for this theoretical treatment. Nevertheless, from the above equations we can determine the character of the dependence of the fraction Θ of the occupied surface and magnitude Φ (determined from (5.66) on the adsorption energy, and the dependence of parameters p and P_B on the amount of adsorbed polymer. Such dependences are shown in Figure 85 for adsorption from dilute solutions.

An analysis of these dependences leads to the following conclusions: the amount adsorbed is independent of the molecular weight and of the equilibrium concentration Φ^*; the adsorbed layer is thin (P_B lies between 2 and 3), and the fraction of the adsorbed segments is high ($p \sim 0.7$), irrespective of the value of x. The value of p is large even when the adsorption energy is low. This means that the adsorption of the polymer is good for all surfaces. The amount adsorbed Γ (in polymer segments per surface site) is of the order of unity. An increase in x leads to adsorption, but if $x > 2 \; kT$, no further improvement in adsorption is observed.

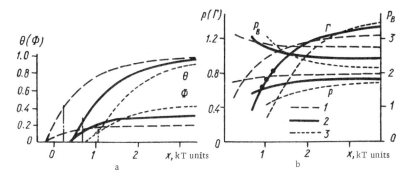

FIGURE 85. Dependence of the degree of surface coverage Θ and Φ (a), and of p, P_B', and Γ (b) on the adsorption energy for different values of X_0 :

1) $X_0 = 1$;   2) $X_0 = 0$;   3) $X_0 = 1$.

The equations derived are considered to different approximations, so that it is possible to find some quantitative correlations between the parameters. If we assume that x is a fixed value (for example, $x = 1$), we can find the values of parameters p and P_B for a given Φ from Figure 86, and calculate Θ and P_S. Then, by inserting these values, we can find the dependence of Γ on Φ at different values of p (Figure 87).

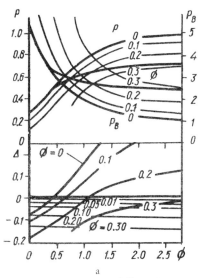

FIGURE 86. Dependence of P_B and p on the adsorption energy at different values of Φ .

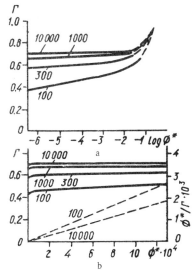

FIGURE 87. Dependence of the adsorption at different degrees of polymerization on log Φ^* (a) and Φ^* (b) at $x = 1$ (dashed line is Langmuir's isotherm).

Figure 87a shows the dependence of Γ on Φ^* (logarithmic scale) over a wide concentrational range, and Figure 87b this dependence for the range of the usual concentrations. The dashed line shows how the values can be fitted on the Langmuir isotherm. Silberberg thinks that under certain conditions his isotherm is identical with the FSE isotherm. However, Frisch et al. do not correlate the shape of the molecule in equilibrium with its behavior during adsorption, and thus *p* is unknown.

The shape of the isotherm predicted by Silberberg (sharp rise and plateau) agrees well with the experimental data. Silberberg's model explains the high *p* values, without postulating high adsorption energies. In general, *p* is little dependent on variations in the given parameters. Evidently, *p* is determined mainly by the structural restrictions on the surface. The model also explains the slow attainment of equilibrium. If, for example, the molecule has 10,000 segments at *p* = 0.5, we need an energy of 3500 kcal to remove a molecule from the surface if the adsorption energy is only 700 cal. This is the energetic barrier that separates the molecules on the surface and in the solution.

Thus, Silberberg's theory leads to the following conclusions.

Adsorption of a polymeric molecule of sufficient size leads to the formation of a surface phase; its composition and structure do not depend on the molecular weight and the concentration in the equilibrium bulk phase.

The concentration of the polymer segments on the surface is actually high, even if the interaction energy of the adsorbent with the adsorbate is low.

The state of the polymeric molecule on the surface is a function of certain specific parameters only; these include the free energy of reaction between the polymeric segments themselves, the polymeric segments and the solution, and the polymeric segments and the surface.

In general, these conclusions agree with many experimental data. It is interesting to note that Silberberg believes that if the thickness of the adsorbed film is a function of the molecular weight, then the proposed model is inapplicable. In such cases either the experimental data do not cover equilibrium conditions, or the system is irreversible.

Silberberg's scheme for the distribution of macromolecules over the surface as a train of segments attached to the surface, and loops stretching into solution, was further developed in a generalized form for any flexible chain adsorbed from a solution /180/. The statistical-mechanical approach was used to calculate the interaction of the polymer with the solvent and of the solvent with the surface, which had not been taken into account in the previous treatment. The partition function of the molecules on the adsorbent surface and in the solution was again determined. If the molecule contains P segments, we can assume that a train of i segments can arrange itself on the surface in $\bar{\omega}\,(i)$ conformations, the loops of i segments can exist in $\omega\,(i)$ conformations and the tail pieces of i segments in $\omega_t\,(i)$ conformations. The molecule can be separated into $\bar{m}_i$ trains, m_i loops, and m_{t_i} tail pieces. The total number of conformations of the system in this case is given by

$$\Omega^S\{\overline{m}_i\ m_i,\ m_{t_i}\} = \left[\prod_i \bar{\omega}\,(i)^{\bar{m}_i}/\bar{m}_i!\right]\left[\prod_i \omega\,(i)^{m_i}/m_i!\right] \times$$
$$\times \left[\prod_i \omega_t\,(i)^{m_{t_i}}\right] \left(\sum_i \overline{m}_i\right)! \left(\sum_i m_i\right)!,$$

(5.77)

where

$$\sum_i i\overline{m}_i + \sum_i im_i + \sum_i im_{t_i} = P,$$ (5.78)

$$\sum_i \overline{m}_i - \sum_i m_i = 1; \quad \sum_i m_{t_i} = 0, 1, 2.$$

Moreover, it is assumed that the difference in the interaction energies between the solvent molecule in contact with the surface and a polymer segment also in contact with the surface is

$$V_{pS} - V_{oS} = \Delta U_S.$$ (5.79)

Then, the canonical partition function of the system is given by

$$q_S(P,\ T) = \sum_{\overline{m}_i, m_i, m_{t_i}} \Omega^S x \exp\left(-\sum_i im_i \Delta U_S / kT\right) \exp\left(-\overline{U}/kT\right),$$ (5.80)

where $\overline{U}$ is the average internal energy of the isolated polymer system in the nonadsorbed state in the solvent. The canonical partition function of the macromolecule in the bulk solvent phase is given by

$$q^B(P,\ T) = \sum_U \Omega(U,\ P) \exp\left(-U/kT\right) \simeq \Omega^B \exp\left(-\overline{U}/kT\right),$$ (5.81)

where Ω_B is the number of conformations associated with energy $\overline{U}$. Hence, if the condition

$$q^S(P,\ T) > q^B(P,\ T)$$ (5.82)

is observed, that is, if

$$\sum_{\overline{m}_i,\ m_i,\ m_{t_i}} (\Omega^S / \Omega^B) \exp\left(-\sum_i i\overline{m}_i \Delta U_S / kT\right) > 1,$$ (5.83)

then the isolated macromolecule will have a lower free energy in the adsorbed state than in the solution.

Thus, we must take into account changes in energy as the result of replacement of the solvent molecules on the surface by polymeric segments. We must find the type of functions $\omega(i)$, $\overline{\omega}(i)$, and $\omega_t(i)$ by random walk techniques, with allowance for the self-exclusion effect. These functions can be represented as

$$\overline{\omega}(i) = \overline{\gamma} i^{-\overline{a}} (\overline{u})^i,$$

$$\omega(i) = \gamma i^{-a} (u)^i,$$ (5.84)

for the number of conformations in two (on the surface) and three (in the bulk) measurements. In these equations $\overline{\gamma}$ and γ are parameters related to the end effects, $\overline{a}$ and a are constants determined by the lattice spacing, and $\overline{u}$ and u are magnitudes related to the coordination numbers of the lattice (in the ideal case they are equal to this number) and characterize the number of possible distributions of the following segments in the surface lattice for the attached segments ($\bar{u}$), or in the bulk (for loops u). By using these functions, we can calculate q^B and q^S in equations (5.82) and (5.83).

From the calculations of Silberberg it is possible to find another interesting magnitude, namely, the change in the free energy on transition of a segment from a loop (of infinite length) into a train (of infinite length). The value ΔF_S is given by

$$\Delta F_S = -kT \ln (\bar{u}/u) - kT \ln \eta = -kT \ln (\bar{u}/u) + \Delta U_S, \qquad (5.85)$$

or

$$(\Delta F_S/kT = -\ln (\bar{u}/u) \eta,$$

where $\ln \eta = (\Delta U_S/kT)$. We can also calculate the mean length P_S of the train of segments on the surface, and P_B of the loops:

$$P_S = \left(\sum_i^\infty i\bar{m}_i / \sum_i^\infty \bar{m}_i \right) = \left(\sum_i^\infty \bar{\gamma}^{-(\bar{a}-1)}\bar{\xi}^i e^{\lambda_2} \right) / \left(\sum_i^\infty \bar{\gamma} i^{-\bar{a}}\bar{\xi}^i e^{\lambda_2} \right) = (\bar{W}')/(\bar{W}), \quad (5.86)$$

and similarly $P_B = (W')(W)$. Here λ_2 and λ_1 are factors

$$\bar{\xi} = \bar{u} \exp [-(\Delta U_S/kT) + \lambda_1]. \qquad (5.87)$$

It can be proved that the ratio $\bar{u}/u$ is independent of the lattice parameters and is about 0.40. Then, the main parameter characterizing the system will be η, that is, ΔU_S is the change in the internal energy of the system during adsorption, and the combinatorial factor $\gamma\bar{\gamma}$ (see (5.84)), where γ relates to the loops and $\bar{\gamma}$ to the adsorbed trains. This product determines the probability of the transition of a segment from the surface into the bulk. The probability of such a transition decreases with decrease in $\gamma\bar{\gamma}$. For the cases considered in the paper of Silberberg, γ and $\bar{\gamma}$ are described by

$$\gamma = [(Z - S)/2]^2/(Z - 1),$$
$$\bar{\gamma} = S/(S - 1)^2. \qquad (5.88)$$

At $Z = 12$ and $S = 6$, the factor $\gamma\bar{\gamma} = 0.196$.

From these theoretical equations it is possible to describe some functions which are very important for characterizing adsorption and the structure of the adsorbed film. Thus, Figure 88 shows the dependence of the average loop length on the average train length for an infinitely long chain at different values of $\gamma\bar{\gamma}$ and $(\Delta F_S)/kT$. This function shows that if ΔF_S is more negative, P_S increases and P_B decreases. If $\gamma\bar{\gamma}$ decreases at constant ΔF_S, then P_S and P_B increase. At low values of $\gamma\bar{\gamma}$, even small variations in ΔF_S may lead to large variations in P_B and P_S.

We shall now consider the fraction of segments contacting the surface

$$p = (P_S)/(P_S + P_B). \qquad (5.63)$$

Figure 89 shows that if $\gamma\bar{\gamma} = 10^{-4}$, a small deviation in ΔF_S from zero leads to a rapid increase in p from 0 to 1. If $\gamma\bar{\gamma} = 1$ or 0.1, this transition will be very slow. Above a certain value of ΔF_S, the adsorbed state becomes unstable and $p = 0$.

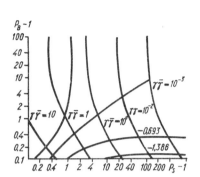

FIGURE 88. Average loop length P_B as a function of the average train length P_S for an infinitely long isolated chain at different values of $(\overline{\gamma\gamma})$ and $(\Delta F_S)/kT$.

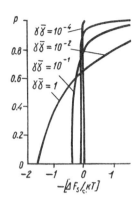

FIGURE 89. Dependence of the fraction of the segments of the isolated chain on $(\Delta F_S)/kT$ at different values of $\overline{\gamma\gamma}$.

From these calculations an interesting fact is also noted, namely that the change in the free energy ΔF_S of the segments on transition from loops on the surface is positive, and that the overall decrement in free energy determining the course of the process is due to the formation of a large number of loops and of segment trains attached to the surface. The more readily the process proceeds, that is, the larger $\overline{\gamma\gamma}$, the higher is the critical value $(\Delta F_S)_{cr}$, corresponding to the beginning of desorption (see Figure 89). Figure 90 shows the number of segments attached to the surface ($\overline{\gamma\gamma} = 0.2$) as a function of $[(\Delta U_S) - (\Delta U_S)_{cr}] kT$ for different values of parameter a in (5.84).

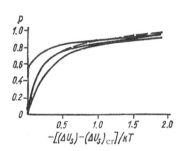

FIGURE 90. Dependence of the fraction of segments attached to the surface on $[(\Delta U_S)-(\Delta U_S)_{cr}]kT$ for different values of a and $\bar{a}$.

The adsorption of a linear flexible chain thus leads to the transition of an isotropic random coil to a molecule, which, according to Silberberg's definition, is a cooperative structure of segment trains (two-dimensional random walk), alternating with loops freely extending into the solution (three-dimensional random walk restricted by barrier). As a result, the molecules adsorbed by the surface generate a phase attached to it. The phase thickness is proportional to P_B, and the concentration Φ_B of the segments in this phase differs appreciably from that in the bulk phase in equilibrium, and from the surface concentration Θ.

In a later work /181/ Silberberg considers the correlation between the parameters of the adsorbed phase Θ and Φ_B, and of the adsorbed molecule P_B, P_S, and p and the concentration in the bulk Φ^*, number of segments in the molecule P (molecular weight), the parameter $\overline{\gamma\gamma}$ describing the flexibility of the chain and its attachment to the surface, interaction parameters polymer − solvent χ and polymer − surface χ_S. The author uses here the quasilattice model of the adsorbed film, which is applied to describe the properties of polymer solutions.

The interaction parameter χ_S is determined as $\Delta U_S = - \chi_s kT$, and accordingly $\eta = \exp(\chi_S)$ and

$$- [\Delta F_S/kT] = \ln(u_S/u) + \chi_S. \qquad (5.89)$$

For the first adsorbed film contacting the adsorbent, a two-dimensional lattice with coordination number C_S, containing M_S sites accessible to the solvent molecule or the segment (provided that the volumes are equal) is considered. Suppose that N_P molecules are adsorbed on the surface. This number of molecules contains PN_P segments, of which M_{SP} are on the surface. In this case the surface coverage is expressed by

$$\Theta = M_{Sp}/M_S. \qquad (5.90)$$

Hence

$$M_{Sp} = \left[\sum_i i m_{Si} \right] N_p \qquad (5.91)$$

segments will be on the surface. In (5.91) m_{Si} is the number of segment trains with number i on the surface, and m_{Bi} is the same parameter in the loops. The number of chain segments in the loops is expressed by

$$M_{Bp} = \left[\sum_i i m_{Bi} \right] N_p = p N_p - M_{Sp}. \qquad (5.92)$$

American scientists /182−185/ considered the problem as to the character of the distribution of segments near the surface. They proved that if the excluded volume is not taken into account, the density of the segments near the surface decreases exponentially with the distance from the surface. Moreover, the characteristic width of the distribution is proportional to the average loop length P_B. However, if the excluded volume is allowed for, the distribution will differ (cell-type) and there will be a sharper transition from the high-density to the low-density region. For the sake of simplicity, it is assumed that the distribution is a step function, and that the volume fraction Φ_B of the polymeric segments is constant in this phase.

We shall denote the film thickness, that is, the number of layers, each th M_S sites, as δP_B^β, and assume that $\beta = 1$ and $\delta = {}^1/_2$. Behind this nogeneous region, termed "volume-surface phase" (subscript B), with tal number of sites $M_B = \delta P_B^\beta M_S$, the true phase of the polymeric tions extends with the volume fraction Φ^*, from which the adsorption th takes place. It is also assumed that the interface is sharp, and that all gments in phase B belong to the adsorbed molecule and are segments arranged in loops. Then $\Phi_B = M_{Bp}/M_B$, and with allowance for (5.90)

$$\Phi_B = (P_B/P_S)\,\Theta\,(1/\delta P_B^\beta), \tag{5.93}$$
$$\Phi_B/(P_B^{1-\beta}\delta^{-1}) = \Theta/P_S. \tag{5.94}$$

This equation expresses the stoichiometric relation between the surface coverage of the volume-surface phase B and the coverage of the surface layer S. If $\beta = \frac{1}{2}$ and $\delta = \frac{1}{2}$, then

$$\Phi_B = 2\Theta/P_S. \tag{5.95}$$

This means that if $\Theta = 1$, that is, the adsorbent is saturated, $P_S \gg 2$. Hence at $\Phi_B \gtrsim \Phi^*$ $P_S \gtrsim 2/\Phi^*$.

The coordination number of the lattice C_B is the same in the surface-volume phase B and in the equilibrium solution. The number of coordination possibilities will be $\frac{1}{2}(C_B - C_S)M_S$ for the attachment of the upper layer S to phase B. The first of these contacts, namely proceeding from M_S surface sites into the volume B, is characterized by the coordination number

$$C_{SB} = \frac{1}{2}l(C_B - C_S). \tag{5.96}$$

The contacts passing from M_B sites in layer B correspond to the effective coordination number

$$C'_{SB} = C_{SB}\,(M_S/M_B), \tag{5.97}$$

and the effective coordination number in phase B will therefore not be C_B but

$$C'_B = C_B - C_{SB}\,(M_S/M_B). \tag{5.98}$$

Such a complex character of the lattice is taken into account in the calculation of the configuration factor and configuration energy, required for calculating the partition function. Besides this complex lattice, three different types of interaction are considered: segment − segment (pp), segment − solvent (po), and solvent − solvent (oo), and the number of contacts determined from the above coordination numbers. To calculate ψ and ψ_S, the concept of the level of the potential interaction energy is introduced for the corresponding types of contacts and for the contacts of types surface − solvent (S_0) and polymer − surface (ps). Silberberg allowed for these parameters, and found the following equations for χ and χ_s :

$$\chi = (C_B/kT)\left[V_{op} - \left(\frac{1}{2}\right)(V_{pp} - V_{oo})\right], \tag{5.99}$$
$$\chi_S = [C_{SB}/kT]\left[V_{So} - (V_{Sp} - V_{pp}) - \frac{1}{2}(V_{oo} - V_{pp})\right]. \tag{5.100}$$

Parameter χ characterizes the interaction energy between polymer and solvent, and χ_S the energy of the exchange of a solvent molecule on the surface by a polymeric segment. If $\chi_S < 0.7$, adsorption is impossible.

Therefore, for good adsorption, the solvent chosen must not strongly solvate the surface.

With allowance for the above factors, it is possible to calculate the partition function for an isolated molecule on the surface $q^S (P, T)$, and for a random macromolecule in solution $q^* (P, T)$. The ratio is given by

$$q^S (P, T)/q^* (P, T) = \alpha^{-P}, \qquad (5.101)$$

where, if P is sufficiently large, α is independent of P and represents the relative change in the activity of the macromolecule when it passes from the bulk of the solution to the surface. Calculation of the partition function yields the following expression for the adsorption isotherm:

$$\ln \alpha = (1/P) [A_0 + \chi B_0], \qquad (5.102)$$

which contains the number of segments in the molecule as one of the parameters. In this equation

$$A_0 = \ln [P_B \Phi^*/(P_B + P_S) \Phi_B] + \ln [(1 - \Phi_B \chi_B)/(1 - \Phi^* \chi^*)], \qquad (5.103)$$

$$B_0 = \frac{2}{C_B} \left[\frac{(1 - \Phi_B)^2}{(1 - \Phi_B \chi_B)^2} - \frac{(1 - \Phi^*)^2}{(1 - \Phi^* \chi^*)^2} \right]. \qquad (5.104)$$

Parameters χ_B and χ^* characterize the number of contacts of different types on the surface and in the bulk.

The adsorption of the polymer is determined not only by the nature of the solvent as such, but also by the energy changes that accompany the exchange of the solvent molecules on the surface by polymer molecules. These changes can be characterized by the energy adsorption parameter $\chi_{P.S}$, introduced earlier. The state of adsorption can be changed by introducing an additive able to displace both polymer and solvent from the surface. This case was theoretically considered by Silberberg /149/. Here also the interaction parameters of Flory – Huggins for polymer and additive χ_{Pl}, solvent and additive χ_{ol}, and additive and surface χ_{lS} are introduced.

Silberberg used a method similar to those in other studies, and obtained an expression for the partition function. He allowed for the introduction of the additive into the solution by parameter ρ, which is the fraction of the adsorption sites occupied by the additive, and concentrations ρ^*, ρ_B, ρ_S of the additive in the bulk of the solution, in layer B and layer S, respectively. The author obtained a complex system of equations, which can be simplified if we consider the case when $\chi_1 = \chi_p$, and $\chi_{ol} = 0$, that is, the solvent and the additive are mixed athermally, and both substances are good solvents for the polymer.

At the same time, there is real competition for the sites on the surface, determined by parameters χ_{PS} and χ_{lS}, while the substitution of the molecules of the additive for the solvent molecules is governed by the equation for the Langmuir isotherms. The presence of the polymer has no effect on the given process, while the presence of the additive affects the polymer adsorption, and is allowed for by χ_{lS} and the additional term χ_{SP} – $- \Delta\chi_{SP} = -\ln [1 + \rho^* (l^* S l - 1)]$, introduced into the appropriate equations.

If χ_{SP} is sufficiently large, then Θ, ρ, and Θ/ρ do not strongly depend on χ_{SP}, as follows from the theory. However, if $\chi_{SP} < 0.7$, desorption takes place.

The region where χ_{SP} strongly affects adsorption will thus be $0.7 < \chi_{SP} < 2.3$. If χ_{SI} has such a value, then with allowance for $\Delta\chi_{SP}$, χ_{SP} falls into this region, and the effect of the additive will be substantial. If χ_{SI} is negative, and $\chi_{SP} < 0.7$, the additive may lead to adsorption of the polymer at sites where it has not been adsorbed before. Finally, if $\chi_{SP} > 0.7$, and χ_{SI} is positive, the polymer may be desorbed.

Besides equation (5.102), another series of equations was obtained by the author, with which it is possible to evaluate the surface coverage Θ, the fraction of attached segments $p = P_S / (P_B + P_S)$, the ratios Θ/p and P_B of the concentration of the solutions, the total number of segments in the molecule (molecular weight), the cooperative factor $\gamma_B \gamma_S$, the interaction parameters χ and χ_S, and some parameters determining the configuration of the polymeric molecules. For two cases, athermal solution ($\chi = 0$) and Θ-solvent ($\chi = 0.5$), Silberberg gave different values to the above parameters, and calculated the theoretical dependences of the adsorption and film thickness, determined by P_B, on the concentration of the solution and the molecular weight. When the calculated curve was compared with the published data, it was seen that the theoretical concepts correctly predict the behavior of real systems.

The calculations indicate that the results for finite concentrations differ appreciably from the values of the adsorption calculated for infinitely dilute solutions. In particular, the film thickness is much larger, and the number of segments on the surface is much smaller, than in the case of the isolated chain. The introduction of parameters χ and χ_S makes it possible to allow for the nature of the solvent, which was determined well by experiment.

The experimental dependences of Θ/p and P_B on the concentration correspond to the dependences predicted by Silberberg's model. From the calculations it follows that at given p and χ the concentration in the surface layer Θ is determined mainly by parameter χ_S, and the concentration in the boundary phase Φ_P is a function of Φ^*. These calculations are interesting because they can be used to theoretically predict the thickness of the adsorbed film, that is, the amount of polymer adsorbed as loops.

Silberberg considers that if the effect of the nature of the solvent and of the concentration is taken into account, there is no need for concepts postulating that the molecule is always sorbed as a random coil, or concepts on multilayer adsorption. However, the author does not deny that multilayer adsorption is possible, especially from concentrated solutions, but he postulates that in this case the second layer of macromolecules condenses on the first layer, and has no direct contacts with the surface.

Silberberg's theory is very useful for an understanding of the adsorption processes of polymers, although it cannot be considered as universal. This is mainly because the theory is based on the definite character of the solution lattice, and thus the same restrictions are imposed as in the theory of polymeric solutions. Therefore, any hypothesis based on a lattice model requires special checking in each individual case. Other hypotheses of Silberberg have been criticized. For example, it has been shown that the idea of a sharply peaked distribution of adsorbed train segments and loops is incorrect /186/.

OTHER THEORIES

The concepts given on the adsorption of polymers are the most developed, and have the greatest theoretical justification, but they by no means cover all the problems involved. When Hoeve /187/ considered the mechanism of adsorption, he took into account that the polymeric molecules readily change their shape under the effect of small interferences. This allowance for configurational changes forms the basis of his theory, but it should be noted that all theories on adsorption more or less allow for configurational changes. However, the forces that act on the surfaces of the adsorbents are usually not accurately known, and therefore cannot be accurately predicted, that is, the molecular theory cannot yet be developed in full detail /187/. It is also interesting to note that the theory of adsorption based on calculation of the configuration of the molecule functioning as a reflecting barrier is incorrect in the case when the chain is not repelled by, but attached to, the surface, that is, adsorbed /187/.

Hoeve et al. /186/ analyze the adsorption of polymers from the point of view of statistical mechanics, that is, on the basis of the above model of alternating trains of adsorbed segments and loops. However, the main emphasis is on the size distribution of the loops. A low surface coverage is considered, when the molecules on the surface do not interact with one another. The behavior of the molecules in the presence of a reflecting barrier is taken into account in the SFE theory, while the attractive forces of the surface are neglected. Therefore, in this theory the number of adsorbed units is proportional to the square root of the chain length, and not to the chain length, as follows from Silberberg's theory. However, Silberberg postulates a sharply peaked distribution of the loops. Hoeve et al. /186/, in contrast to Silberberg, do not postulate a freely linked chain, but allow for the chain stiffness, which leads to larger loops for a flexible polymer and to a low free energy of adsorption. When they solve the problem, they apply a mathematical method similar to that used when considering the helix-coil region of DNA. It is considered that the configuration of the polymeric chain depends on steric hindrances and on attractive forces between the groups along the chain, and on the forces of interaction between chain and surface. In this case the authors assumed the existence of adsorbed trains and loops, applied Gaussian statistics, and calculated the partition function as

$$q_n = \sum (m!)^2 \prod_i \left(\frac{w_i^{n_i}}{n_i!} \right) \prod_i \left(\frac{v_j^{m_j}}{m_j!} \right), \qquad (5.105)$$

where $m = \sum_i n_i = \sum_i m_i$, that is, the number of loops and trains, equal to m_i and n_i, relates to the segments in the trains (m) and in the loops (n). There are $n_1, n_2, n_3, ..., n_i$ loops containing 1, 2, . . ., i nonadsorbed units, and $m_1, m_2, ..., m_j$ trains containing 1, 2, . . ., j units. In (5.105), w_i and v_j are partition functions for a loop of size i and a train of size j. When the values of w_i and v_j are calculated, the formation of loops between two trains requires a different change in the chain conformation (bending away), and the contribution of such bending to the partition function will be greater if the number of conformations of approximately equal energy available to the chain is larger, and the difference in the rotational angles corresponding to these conformations is greater.

Hoeve et al. /186/ consider two cases: a flexible and a stiff chain. The partition function of the loops cannot be equal to that of the free chain, since the loops are attached to the surface at their ends. From these physical postulates, the authors obtained the equation

$$w_i = c_i^{-\frac{3}{2}}, \qquad (5.106)$$

where c is a factor for the combined effects of chain flexibility (for a flexible chain $c = 1$, for a stiff one $c = 0$); in addition, $v_j = \sigma^j$, where σ is the ratio of the partition function of the adsorbed unit to that of a unit in the bulk, which characterizes the attraction of the segments by the surface.

The partition function of one molecule is

$$q_n = \sum (m!)^2 \prod_j \left(\frac{\sigma^{jm_j}}{m_j!} \right) \prod_i \frac{\left(c_i^{-\frac{3}{2}} \right)^{n_i}}{n_i!}. \qquad (5.107)$$

By applying the standard method of the Lagrangian multiplier, we obtain

$$n_i = m c_i^{-\frac{3}{2}} e^{i\lambda} e, \qquad (5.108)$$
$$m_j = m \sigma^j e^{i\lambda} e^{-\xi}, \qquad (5.109)$$

where λ and ξ are the multipliers. The final result of the calculations is as follows:
partition function of one molecule,

$$q_n = \sum (m!)^2 \prod_j \left(\frac{\sigma^{jm_j}}{m_j!} \right) \prod_i \frac{\left(c_i^{-\frac{3}{2}} \right)^{n_i}}{n_i!}; \qquad (5.110)$$

fraction of attached units,

$$p = \sum jm_j/n = \left(1 + S_{-\frac{3}{2}} \right) \left(1 + S_{-\frac{1}{2}} + S_{-\frac{3}{2}} \right)^{-1}; \qquad (5.111)$$

average loop size,

$$\sum in_i / \sum n_i = S_{-\frac{1}{2}} / S_{-\frac{3}{2}}; \qquad (5.112)$$

average train size,

$$\sum jm_j / \sum mj = S^{-1}_{-\frac{3}{2}} \left(1 + S_{-\frac{3}{2}} \right), \qquad (5.113)$$

where

$$S_{-\frac{1}{2}} = \sum ci^{-\frac{1}{2}} \exp(i\lambda),$$
(5.114)

$$S_{-\frac{3}{2}} = \sum ci^{-\frac{3}{2}} \exp(i\lambda).$$
(5.115)

From these equations the authors derived the adsorption isotherms for a low surface coverage. The molecule is considered to be adsorbed if at least one segment is within a distance δ from the surface, where δ is of the order of the thickness of the segments. If the system contains N molecules, N_n adsorbed molecules, and N_f molecules in the bulk, and finally $N_n + N_f = N$, the partition function of the whole system is

$$Q = \sum [q_n^{N_n}(A\delta)^{N_n}/N_r!](V^{N_f}/N_f!),$$
(5.116)

where A is the surface area of the adsorbent, and V is the volume of the solution. Summation is over all possible values of N_n and N_f. By maximizing (5.116), we obtain

$$N_n/A\delta = q_n(N_f/V),$$
(5.117)

which may also be represented as

$$N_n/A\delta = \exp(-\lambda_n)(N_f/V).$$
(5.118)

This isotherm predicts the dependence of the initial slope on the molecular weight. The equations showed that the size distribution of trains and loops are broad. From the theory it is possible to calculate the dependences of p, m/n, and the free energy of adsorption on c and σ. A change in the nature of the solvent or the surface affects σ, but naturally does not affect c, while temperature variations may affect both parameters. Figure 91 shows the dependence of attached units on σ for different values of c. For a stiff chain $c = 0$, and the curve is characteristic of a transition of the first kind, while at $c > 0$ the transition is diffusional.

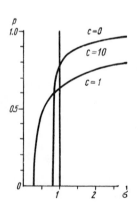

FIGURE 91. Dependence of the fraction of attached segments p on σ for various values of chain flexibility.

On the basis of this theory we can thus predict the behavior of molecules of chains of various flexibilities. For a stiff chain the theory predicts a small number of loops, and a greater fraction of the segments attached to the surface. In this case, adsorption can formally be considered as phase transition. The chain is either adsorbed by all segments, or not adsorbed at all. In contrast to Silberberg, Hoeve predicts a dependence of loop sizes on adsorption energy for flexible chains (large loops at low energies and small loops at high energies).

The ability to form a considerable number of loops is thus directly dependent on the flexibility of the chains and their size, while the number is determined by the free energy of adsorption. The main advantage of this theory is that it allows for chain flexibility.

The theory of polymer adsorption was further developed by allowing for the interaction of the polymeric chains near the surface /185/. The first layer on the surface is considered to be a polymeric solution, described by the Flory — Huggins theory, while the interaction in other layers is assumed to be absent if the polymer is adsorbed from a Θ-solvent. By applying the same mathematical calculations as in /186/, but allowing for chain inter- action, an expression for the partition function can be found. The region near the surface is divided into two subregions: the first layer of thickness σ corresponds to the effective thickness of the segments, and the next layer (loop layer) in which the segments are not directly attached to the surface. In the first layer, with a larger number of adsorbed molecules, the interaction of the segments as the result of this higher concentration will be stronger.

We shall assume such a structure of the adsorbed film, and find the free interaction energy in the first layer. The excess free energy in the first layer, determined by the interaction of segments, can be represented by

$$\Delta F_1/kT = (A\delta/V_1)[(1 - v_2)\ln(1 - v_2) + v_2(1 - \chi_1 v_2)], \qquad (5.119)$$

where V_1 is the volume of the segment; χ_1 is the interaction parameter; v_2 is the volume fraction of the polymer in the first layer, expressed as $v_2 = n_2 x V_1/A\delta$; x is the number of segments in the chain; n_2 is the number of sorbed molecules in the first layer. The total free energy of the adsorbed layer consists of the free interaction energy and the free adsorption energy ΔF_2, calculated for the case when the interaction between the segments is neglected

$$- \Delta F_2/kT = \ln q_n = n(-\lambda - p\eta), \qquad (5.120)$$

where q_n is the partition function for one molecule; n is the total number of segments; p is the fraction of sorbed segments; χ and η are the Lagrangian multipliers.

If N_p molecules are sorbed, the free energy of adsorption will be

$$\Delta F_1 + \Delta F_2 = \Delta F, \qquad (5.121)$$

$$\Delta F/kT = nN_p(\lambda + p\eta) + (A\delta/V_1)[(1 - v_2)\ln(1 - v_2) + \\ + v_2(1 - \chi_1 v_2)], \qquad (5.122)$$

where

$$npN_pV_1/A\delta = v_2. \qquad (5.123)$$

Factor η is expressed in terms of the following magnitudes:

$$\eta = -\lambda - \ln\sigma - \ln\left(1 + S_{-\frac{3}{2}}\right), \qquad (5.124)$$

while

$$p = \left(1 + S_{-\frac{3}{2}}\right)\left(1 + S_{-\frac{1}{2}} + S_{-\frac{3}{2}}\right)^{-1}.$$ (5.125)

Equations (5.123)–(5.125) determine the values of λ, v_2, and p for given N_p and χ_1. The partition function of a system consisting of N_p adsorbed and N_l free molecules is

$$Q = N! \sum [\exp(-\Delta F/kT)(A\delta)^{N_p}/N_p!] \ (V^{N_l}/N_l!),$$ (5.126)

where V is the volume of the solution. From this equation we obtain the equation for the adsorption isotherm

$$N_p/A\delta = (N_f/V)\exp(-\lambda n),$$ (5.127)

similar to (5.118), but λ is a parameter depending on the amount of the adsorbed polymer, according to (5.124).

Figure 92 shows some isotherms for different numbers of segments in the chain (different molecular weights) and values $c = 0.1$, $\sigma = 1$, and $\chi_1 = 0.5$ (Θ-point). The amount of the adsorbed polymer is expressed by the dimensionless magnitude $nN_pV_1/A\delta$, and the concentration of the solution by the dimensionless parameter $nN + V_1/V$ (volume fraction of polymer in solution). These isotherms show that even at a low concentration of the solution, the adsorption increases with increase in molecular weight. It is noteworthy that we cannot use the same value of χ_1 for the first layer and for the solution. It can be shown that the second virial coefficient in the adsorption layer will differ from that in the solution, while for the adsorption layer we must take χ_1 as less than 0.5 at the Θ-point.

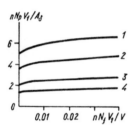

FIGURE 92. Adsorption isotherms for a range of molecular weights:

1) $n = 3.24 \cdot 10^4$; 2) $n = 1.44 \cdot 10^4$; 3) $n = 3.6 \cdot 10^3$; 4) $n = 9 \cdot 10^2$.

In later papers /187, 188/ the theory was made more accurate by allowing for the excluded volume effect. This leads to a decrease in the number of possible chain conformations in the boundary layers. Hoeve assumed, in analogy with Flory's theory, that when the excluded volume is taken into account, the chains are expanded by a factor α. Then the elastic free energy, determined by the expansion of the loop, will be given by

$$\Delta F_{el} = mN_pkT\left[\left(\frac{3}{2}\right)(\alpha^2 - 1) - 3\ln\alpha\right].$$ (5.128)

For small values of $(\alpha - 1)$,

$$\Delta F_{el} = 3mN_pnT(\alpha - 1)^2.$$ (5.129)

In this case the elastic free energy of interaction in the loop

$$\Delta F_3 = \Delta F_{el} + \left(\frac{1}{2}\right) n N_p kT \left(\frac{1}{2} - \chi_1\right) \cdot K\alpha^{-1} v_2(0), \quad (5.130)$$

where K is a theoretical factor, including the layer thickness δ and parameter c; $v_2(0) = npNpV_1/A\delta$.

Hoeve allowed for the elastic free energy, and found the partition function for the canonical ensemble for the total system of N polymeric molecules, consisting of N_S molecules in the solution of volume V, and N_p adsorbed molecules. The adsorption equation found has the form

$$N_p/A\delta = (N_S/V)\exp\left\{-n\left[\lambda + \left(\frac{1}{2}\right)\left(\frac{1}{2} - \chi_1\right)Kv_2\right]\right\}. \quad (5.131)$$

According to this equation, the adsorption is proportional to the concentration of the solution, as in the case of chains which do not interact. For higher adsorption v_2 increases, while p and $(-\lambda)$ decrease. If $\lambda + \frac{1}{2}(\frac{1}{2}-\chi_1) Kv_2 = 0$, adsorption reaches its limiting value at $n \to \infty$.

The theoretical equations show that for a poor solvent (χ_1 is large), the limiting value of $(-\lambda)$ is small, and the size of the loops and the amount of the adsorbed polymer are large. At $\chi_1 = \frac{1}{2}$, interaction in the loops ceases; $(-\lambda)$ decreases to zero when $n \to \infty$, and adsorption becomes unlimited with increase in molecular weight. According to Flory's theory, this is the point of initial separation of the phases, and it can be considered as the beginning of multilayer adsorption. A comparison of the theory of Hoeve /184—187/ with that of Silberberg shows that several parameters which Silberberg considered to be constant in his calculations are variable. Therefore, the numerical results of both theories cannot be compared. However, the general adsorption patterns are similar in both theories.

Not one of the analyzed theories allows for the nature of the adsorbent. However, if the adsorbing surface is characterized by inhomogeneous distribution of the active centers, some of the segments will be attracted and some will be repelled by the surface. These effects are not taken into account in the theory, and may explain some of the experimental results /62/. A quantitative checking of all the theories requires a knowledge of the parameters entering the equation, for example σ, c, and other parameters, which can be found experimentally.

In 1969, El'tekov /156/ proposed another equation for the adsorption isotherm. The author considers that for strong physical adsorption the fraction of nonadsorbed segments is close to zero, and therefore it is sufficient to consider monolayer adsorption with an orientation of the adsorbed molecules of linear polymers parallel to the surface. (We should note that this assumption contradicts the experimentally found p value, and the concepts of loops and of the distribution of the densities of the segments according to distance from the surface.)

To a first approximation, El'tekov assumes strong specific adsorption of the polymer, while the adsorption of the solvent molecules is weak and nonspecific. The equation is derived from considerations of the quasi-reaction of displacement of the solvent molecules M by polymer segments Π from the adsorbent surface

$$\Pi^e + rM^S \gtrless \Pi^S + rM^e, \tag{5.132}$$

where superscript e relates to the solution, and superscript S to the adsorption layer. By expressing the equilibrium constant K for such a system in terms of the activity of the components and the chemical potentials of the polymer on the basis of Flory's theory for athermal polymeric solutions, the author obtained the adsorption isotherm in the form

$$\frac{1}{2} en\,(\Phi_1^S/\Phi_2^S) - \ln(\Phi_2^S/\Phi_2^e) = \ln K, \tag{5.133}$$

where Φ_1^S and Φ_2^S are the volume fractions of the components in the adsorption layer, and K is the equilibrium constant for the given reaction. The values Φ_1^e and Φ_2^e are determined during the analysis of equilibrium solutions. To calculate the equilibrium constant, the dependence of Φ_1^S (volume fraction of polymer in the adsorption layer) on Φ_1^e (volume fraction of polymer in the solution) must be found. El'tekov postulates that in the region of the plateau on the isotherm, the adsorption layer is completely saturated by macromolecules and does not contain the solution. In this case

$$\Phi_1^S = \Gamma_1/\Gamma_{max}, \tag{5.134}$$

where $\Gamma_1 = \dfrac{(\Phi_1^0 - \Phi_1)\,V_{1,2}}{m_a \cdot S}$, $V_{1,2}$, m_a, and S are the volume of the solution, mass, and specific surface of the solid body; Γ_1 is the excess (Gibbs) adsorption; and Γ_{max} is the limiting adsorption.

El'tekov's isotherm was proposed for the case of specific adsorption, but it was tested on the example of the adsorption of polystyrene on carbon black and Aerosil silica where, according to El'tekov, no specific adsorption is observed. The equation does not take into account the flexibility of the chain or its molecular weight, and is very restricted in character.

Pouchly /190/ proposed a theoretical model for the equilibrium behavior of flexible macromolecules in the pores of adsorbents. Its basis is to ascribe the decrease in the number of possible configurations of the chains to geometrical restrictions imposed by the pore walls and the effect of adsorption forces. Hoffman et al. calculated the thermodynamic properties of the adsorbed polymers /191/ on the basis of the theory developed by one of the authors /192/.

* *
*

An analysis of all the theories of polymer adsorption shows that the application of contemporary ideas on the structure and properties of polymers and methods of statistical mechanics have appreciably helped in the understanding and predicting of the adsorptional behavior of polymers. However, no universal equation of the adsorption isotherm can be derived.

This indicates the complex and many-sided phenomena accompanying adsorption. Further research must be undertaken in this field. It is evident that the theoretical concepts on adsorption and the closely related concepts on the structure of the adsorption layer that have been developed must be taken into account when analyzing the properties of adsorbed films and surface layers of polymers on the interface, and also when elaborating an adsorptional theory of adhesion, reinforcement, etc.

We should note that the adsorbed films and the conformations of the molecules near the interface have been statistically examined in several papers /131, 146, 154, 165, 182—190, 193/.

Chapter 6

SPECIAL FEATURES OF ADSORPTION
FROM CONCENTRATED SOLUTIONS

The adsorption of polymers on solid surfaces is theoretically considered in most papers on the basis of the theory of dilute solutions, which has been elaborated in detail and can be used for studying adsorption. This refers mainly to problems involving the conformations of adsorbed chains, and the nature of their attachment to the adsorbent surface. Therefore, most of the experimental papers deal with adsorption from dilute solutions.

However, in practice, polymers are frequently adsorbed from concentrated solutions (application of glues, coatings, etc.). The pattern of such adsorption should be different from adsorption patterns from dilute solutions.

Concentrated solutions are those in which the random coils of the macromolecules begin to overlap, so that the coils are compressed and their size decreases /194/. Compression begins at a concentration of the solution approximately equal to the inverse of the intrinsic viscosity.

A concentrated solution can also be defined in another way: the mean concentration of the polymer in it is equivalent to the mean concentration of the polymeric segments in the coils.

The appearance of molecular interaction forces in solutions leads to the appearance of molecular aggregates or other supermolecular structures, in which the conformation of the macromolecules differs from that in dilute solutions. From studies on concentrated polymer solutions it was concluded /195/ that with increase in the concentration of the solution, transition from coiled conformations to uncoiled ones typical of dilute solutions is possible, because of the redistribution of intermolecular forces. Such a process occurs in a (thermodynamically) poor solvent, when the decrease in the coil size during its overlapping may lead to a transition of the chain into a less coiled conformation, which as the result of stronger intermolecular interaction is thermodynamically more favorable.

As the concentration of the polymer solution increases, the process of cross-linking and of the formation of a fluctuational thixotropic or constant structural network takes place with increasing intensity so that gelation is possible /196/. The structure of a polymeric gel, as assumed in these papers and later confirmed by electron microscope studies /197/, is formed because of the appearance of a continuous network, penetrating through the whole bulk. Such a network is not very probable for coiled molecules.

Changes in the chain conformation with deterioration in the solvent power may thus differ in dilute and concentrated solutions. The strengthening of intermolecular reaction in a concentrated solution, and the appearance of molecular aggregates, lead to changes in chain conformations when the external conditions are varied and differ from those in dilute solutions. Therefore, if we allow for the contributions of molecular interaction to the polymer structure, we can consider that the concepts of the behavior of a macromolecule in dilute solution cannot be applied to the properties of a concentrated solution in which a strong intermolecular reaction takes place. This must be borne in mind even when considering data of polymer adsorption by solid surfaces from moderately concentrated solutions.

In fact, according to Perkel and Ullman /54/, adsorption does not reach saturation when the concentration of the solution is increased. When the adsorption of elastomers on carbon blacks from different solvents (the elastomer concentration was varied between 0.01 and 0.2%) was studied, very characteristic adsorption isotherms were obtained, with a maximum at some concentrations. However, from sufficiently concentrated solutions (26 mg/ml) such adsorption isotherms do not reach saturation, and adsorption increases with increase in concentration /88, 89, 198/. This fact was not convincingly explained at that time.

Since macromolecules change their conformations, and cross-linking occurs in concentrated solutions, it would be natural to correlate the anomalies observed with cross-linking in the solutions. In other words, we should consider the effect of the solution structure on adsorption, and the polymer should be adsorbed from solutions that are more concentrated than those used in most studies /195/.

We first investigated the dependence of adsorption on the degree of cross-linking of the solution /199—202/. Since the character of structure formation in concentrated solutions is influenced by the nature of the solvent, we studied the effect of solvent power, temperature, and concentration of the solution, on adsorption. The results may give information on the character of polymer — surface interaction in a concentrated solution, which is the first act in the formation of the polymer film on the surface, while the structure and physical properties of the filled polymer depend on the conditions of its formation. Adsorption interaction between polymer and filler takes place in solution, and influences the properties of the surface films even when the solvent is absent.

To explain the influence of the above factors, we studied the adsorption of some polymers of different chemical structure. Glass fibers, which are the most important fillers for reinforced plastics and filled polymers, were used as the adsorbent. The polymers used were poly (methyl methacrylate), polystyrene, poly (methacrylic acid), and copolymers of methacrylic acid with styrene or gelatin. Solvents of different powers were employed.

The concentrational range was up to 2—2.5%, that is, in these solutions intermolecular reaction can no longer be neglected. Figure 93 shows typical isotherms of the adsorption of polystyrene from a good (benzene) and poor (cyclohexanone) solvent.

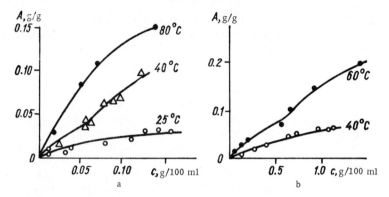

FIGURE 93. Isotherms of the adsorption of polystyrene by glass fibers from benzene (a) and cyclohexanone (b).

The following tests were carried out in studies on the effect of the surface state on adsorption. The glass fibers contained Mn. By treating the fibers with HCl, Mn could be removed. The fibers became lighter and less strong because pores were formed, but this did not lead to any appreciable differences in adsorption.

Figure 94 shows the isotherms of the adsorption of poly (methyl methacrylate) from various solvents. It can be seen that the curves pass through a maximum.

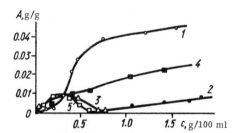

FIGURE 94. Isotherms of the adsorption of poly (methyl methacrylate) at 19°C:

1) acetone; 2) acetone + 5% ethanol; 3) toluene;
4) acetone + 5% heptane; 5) toluene at 25°C.

We studied the adsorption of poly (methacrylic acid) from aqueous solutions in the concentrational range from 0.06 to 2 g/100 ml, and from organic solvents of the same concentration. However, when the polyacids were neutralized to 5, 10, 15, and 50%, no adsorption was observed in any of these cases.

In the adsorption of the copolymer of styrene and methacrylic acid (the copolymer contained 1.6% mers of the acid) from benzene and cyclo-hexanone, it was shown that the copolymer was not adsorbed from

cyclohexanone at room temperature, as in the case of pure polystyrene at
room temperature.

We also studied the adsorption of the copolymer containing 26% poly-
(methacrylic acid) mers from an acetone and benzene solution (volume
ratio 1 : 1), and from dioxane. No adsorption from these solvents was
observed, but the copolymer does not dissolve in other solvents.

Gelatin is only slightly adsorbed from aqueous solutions at 30°C, in
contrast to the adsorption of the other polymers that we studied. For
gelatin, as for poly(methyl methacrylate), the curves pass through a
maximum at concentrations of about 0.2%, while at a concentration of 0.4%
no adsorption was observed. The adsorption of gelatin from aqueous
solutions in the presence of urea is somewhat weaker, and the maximum of
the adsorption that remains decreases, and is displaced to the right-hand
side at higher concentrations.

When porous glass fibers (the initial fibers were treated with HCl) were
used, there were no great changes in adsorption.

We also studied desorption, and proved that after adsorption from
moderately concentrated solutions, desorption proceeds rapidly and
almost completely.

We shall now discuss our results. It is interesting that the values of the
adsorption found are much higher than those from dilute solutions.

The adsorption of polystyrene from a good solvent (benzene) is better
than from a poor one (cyclohexanone). For poly(methyl methacrylate),
when solvents of different power are used, the adsorption and the form of
the isotherm are greatly influenced. The isotherms of the adsorption of
poly(methyl methacrylate) from acetone or chloroform solutions are
flatter than those of polystyrene from different solvents, but the adsorption
isotherms of poly(methyl methacrylate) from toluene are curves with a
sharp maximum near 0.3 g/100 ml.

The higher values of adsorption and the unusual shape of the isotherm in
some cases indicate the complex character of adsorption from moderately
concentrated solutions. To explain these patterns, we cannot completely
apply the idea developed for the adsorption of low-molecular weight
compounds. It is evident that adsorption does not lead to the formation of
a monomolecular layer on the surface of the adsorbent, since the values
of adsorption are too high. It appears that when adsorption from moderately
concentrated solutions is considered, we must allow for the existence of
supermolecular structures of molecular aggregates in solution. The
higher values of adsorption can be explained by the transition of aggregates
onto the adsorbent surface instead of isolated macromolecules, as in the
case of dilute solutions. The character and degree of cross-linking will
determine the adsorption.

In a good solvent, where as the result of weaker molecular interaction
there is less cross-linking of the solution, adsorption is higher than in a
poor solvent, where because of interaction between the molecules and their
aggregates transition of the polymeric molecules to the surface is more
difficult. It is possible that it is this phenomenon that leads to the absence
of adsorption of polystyrene from cyclohexanone solution at 25°C, although
adsorption is observed from a good solvent.

Temperature increases may lead to either improvement or deterioration of the solvent. Thus, in good solvents the intrinsic viscosity decreases with temperature, and hence the power of the solvent decreases. This leads to a growth of molecular aggregates, and increase in the adsorption of molecules from a good solvent. However, the intrinsic viscosity increases with temperature in poor solvents, that is, the solvent deteriorates. This leads to weaker interaction between the aggregates and thus to an increase in adsorption (poly (methyl methacrylate) — acetone).

Higher temperatures may, however, lead to a deterioration in the solvent (polystyrene — cyclohexanone), and to growth in the chain aggregates.

Our data show that during the adsorption of the polymer from moderately concentrated solutions, higher values of adsorption are observed from these solvents, where the degree of cross-linking is smaller, or where the conformation of the chains is more elongated, in other words, from good solvents. According to Schulz and Kantow /105/ acetone is a somewhat poorer solvent for poly (methyl methacrylate) than is toluene. However, the more elongated form of the chains in toluene (relatively poor solvent) apparently affects the structure formation more, so that adsorption decreases.

We know that in a poor solvent the solution is cross-linked earlier if the polymeric chain is more elongated. Apparently, in toluene solutions cross-linking starts at lower concentrations than in acetone, and thus the adsorption from these solutions is lower than that from acetone solutions.

The increase in adsorption from acetone solutions with increase in concentration can be explained by an increase in the size of the molecular aggregates which pass onto the adsorbent surface. Thus, in this case the higher adsorptions are explained by the transition of molecular aggregates, and not of isolated coils, onto the surface. The degree of molecular aggregation may increase with increase in concentration and with deterioration in the solvent. Therefore, at low concentrations of the solution we observe higher adsorption from those solvents in which these structures are formed earlier. With increase in the concentration, the mutual interaction of the macromolecules and supermolecular structures becomes stronger, and leads to the appearance of a continuous spatial network of the solution. Hence the macromolecule can no longer pass onto the surface, and the adsorption passes through a peak and then falls to zero.

This might also explain the similar course of the adsorption of gelatin solutions. The introduction of urea into the solution hinders molecular interaction in the solution. Thus, the size of the aggregates and adsorption at the maximum of the isotherm decrease. The maximum shifts to the right, since in the presence of urea structure formation takes place at higher concentrations.

Later, we obtained isotherms with a maximum for the polymers (polyaminostyrene) adsorbed by NH_4Cl from dichloroethane (Figure 95), and for oligomers (oligoethylene glycol adipate) adsorbed by the same adsorbent from the same solvent. During the last years the more complex dependences of adsorption on concentration have been observed in studies on the adsorption of polymers and oligomers on solid surfaces. Thus, Ermilov /93/ studied the adsorption of PF-6 lacquer and glyptal lacquer on rutile and iron minium, respectively, and obtained isotherms with two maxima at different equilibrium concentrations. The isotherms of the adsorption of polydimethylsiloxane on Aerosil silica are also complex /118/.

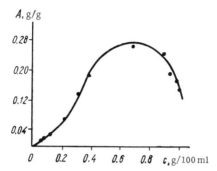

FIGURE 95. Isotherms of the adsorption of poly-aminostyrene by ammonium chloride from dichloroethane solution.

Isotherms with maxima are frequently observed in the adsorption of oligomers on solid surfaces, for example, in the adsorption of oligoesters of adipic acid and tert-butylene glycol on titanium dioxide and ferric oxide from benzene solutions /205/. The adsorption of three series of oligomers was studied. The molecules of the first series contained the COOH groups at the ends, the molecules of the second series contained COOH and OH groups, while the molecules of the third series contained OH groups only. In all cases the isotherms were obtained with maxima at different concentrations of the oligoesters and oligoethers. The absolute magnitude of the adsorption depends on the character of the terminal groups.

The isotherms of the adsorption on titanium dioxide of oligoesters based on ethylene glycol and succinic, adipic, or sebacic acid are similar in shape: all the isotherms pass through a peak, and then fall more or less sharply /206/.

The complex character of the isotherms was shown by Akhmedov and his co-workers when they studied the adsorption of polyelectrolytes on different adsorbents (clays, glass fibers, etc.) /207, 208/. Trapeznikov /215/ found that in the adsorption of polydimethylsiloxane on powdered silica gel the isotherm passes through a maximum at a concentration of 5 g/100 ml.

The degree of cross-linking of the solution, the size of the molecular aggregates, and the elongation of the macromolecules, all affect adsorption from moderately concentrated solutions. The increase in the adsorption of poly(methyl methacrylate) from acetone solutions with increase in temperature can be explained by the decrease in the degree of cross-linking of the solution, and transition of the molecules into a less coiled conformation. In adsorption from toluene, the lower degree of cross-linking at higher temperatures extends the concentrational range in which the polymer is adsorbed. At the same time this leads to a decrease in the value of the adsorption at the maximum of the isotherms because the macromolecular coils and aggregates become smaller.

Data on the adsorption of polymers in the presence of a precipitating agent indicate that if the agent does not cause structure formation, then displacement of the molecules in the solution leads to a strong decrease in their adsorbability. Besides structure formation, other factors must be taken into account, such as blocking of active groups on the adsorbent surface because of their interaction with the solvent molecules. Thus, in the adsorption of gelatin on glass from aqueous solutions, the $Si - OH$ groups of the adsorbent appreciably interact with water, which greatly decreases the adsorption of the gelatin molecules.

The shape of the macromolecule strongly affects adsorption. In the case of poly (methacrylic acid), which in the range of concentrations studied has globular macromolecules, no adsorption takes place. The presence of globular structures reduces possible contacts of the molecule with the adsorbent surface, which in turn strongly reduces the strength of the possible bond with the surface and leads to preferential adsorption of the solvent. The formation of an uncoiled molecule should increase interaction with the surface /209/. Thus, adsorption is observed for the more uncoiled gelatin molecules. However, uncoiling of the poly (methacrylic acid) chains because of neutralization does not increase adsorption, which might be the result of deterioration in the conditions for chain aggregation in solutions of comparatively low concentration (not more than 2%), and of stronger hydration of the COOH groups /210/. Data on the adsorption of the copolymer of styrene and methacrylic acid can be explained in a similar manner. The copolymer with a higher content of carboxyl groups probably forms fairly stable globular structures in the solution because of intra-molecular reaction, and so no adsorption takes place. For the copolymer with a small number of COOH groups, adsorption proceeds well from a good solvent, but not from a poor one.

During the adsorption of supermolecular structures, transition onto the adsorbent surface should lead to a decrease in the strength of the bond with the surface with increase in concentration. We believe that this explains the complete desorption of the polymer from the adsorbent surface when this surface is treated with the solvent.

We thus conclude that the character of polymer adsorption from moderately concentrated solutions depends also on other factors, mainly on the shape of the macromolecules in solution, and on the degree of cross-linking of the solution. Stronger molecular interactions in the solution should increase adsorption up to a certain limit, because the size of the molecular aggregates passing onto the surface increases. Further increase in the cross-linking of the solution may hinder such a transition, and thus reduce sorption.

In this case the decisive factor in adsorption is the specification of the formation of supermolecular structures in the solution, and of a spatial network of the solution. Thus, we assume that the adsorbed film is formed by aggregates of polymeric molecules, and possibly the adsorption of polymers on the surface of a solid substance can be considered as mono-molecular with respect to this substance.

The hypothesis on the transition of molecular aggregates onto the surface was experimentally confirmed when the adsorption of oligoethylene glycol adipate (molecular weight 2000) on the surface of Aerosil silica and carbon black was studied in good (acetone) and poor (toluene) solvents. The turbidity spectra were investigated /58, 59/.

It was found that molecular aggregates are present in oligoethylene glycol adipate solutions. The size of the aggregates depends on the concentration of the solution and the solvent power (Table 21). The number of such aggregates changes from $0.038 \cdot 10^8$ to $30 \cdot 10^8$ in unit volume. With increase in the concentration of the solution, the size of the aggregates increases at first insignificantly, but when the concentration exceeds 20 mg/ml, the increase is greater. At higher concentrations (80—100 mg/ml), the size of the aggregates no longer grows. The maximum size in acetone is $0.13\,\mu$.

TABLE 21. Size of aggregates in solutions after adsorption

Concentration of solutions, g/100 ml	Time, hr	Size, A	n
	Acetone		
10.00	0	<100	6.40
	4	<100	5.10
	24	180	3.95
	48	300	3.85
	72	600	3.35
	336	1200	2.50
	Toluene		
3.00	0	100	5.06
	4	100	4.31
	24	300	3.85
	48	3200	2.08
	72	6000	1.96

Note. n is the slope of the curve of the extinction plotted against the wavelength.

The solvent power estimated from the intrinsic viscosity strongly affects the size of the aggregates. In a good solvent, for example, acetone ($[\eta] = 0.175$), the size of the aggregates increases to $0.08 - 0.70\,\mu$. The formation of molecular aggregates was experimentally confirmed even in oligomers. To check whether these aggregates pass onto the adsorbent surface during adsorption, the turbidity of the solution was measured before and after adsorption. The adsorption was carried out under static conditions in the concentrational range of 25—1000 and 25—30 mg/ml for acetone and toluene, respectively. The kinetic curves shown in Figure 15 are somewhat unconventional. During the first minutes after the adsorbent is mixed with the solution, the adsorption is stronger than after a certain period has elapsed. This is apparently due to redistribution of the adsorbed macromolecules and their aggregates over the surface.

It was found that adsorption from a poor solvent is higher than from a good one. The isotherms have adsorption peaks. When the dependence of the extinction of the solution on the wavelength was studied, it was found that the molecular aggregates disappear during the adsorption process.

This confirms the assumption that during adsorption molecular aggregates become adsorbed on the surface. Since in the solution the molecules bound into aggregates and the unbound molecules should be in equilibrium, this state is disturbed after adsorption.

The turbidity spectra of the solutions show that in a day or two (depending on the concentration) the aggregates are reformed. This is indicated by the decrease in the coefficient n, which is inversely proportional to the size of the aggregates. The size of the aggregates increases more rapidly in toluene than in acetone, which is due to the thermodynamic activity of the solvent. The results of these studies are shown in Table 21. The transition of the molecular aggregates onto the surface, proved experimentally, can be used to explain both the high adsorption of relatively low-molecular weight oligomers and the concentrational dependence of adsorption (isotherms with a peak). However, further research is required to explain the preferential adsorption of aggregates.

From these data we can conclude that deviations in the course of adsorption from the theoretically predicted behavior will be greater at higher concentrations of the solution. The research of Lipatov /195/ shows that one of the reasons for the poor agreement between the theoretical equation and the experimental data is that it is impossible to allow for the real structure of the polymer solutions. All the theories are based on the concept of the behavior of macromolecules in solutions of limiting dilution. However, in real systems we deal with finite concentrations of solutions, where molecular interaction can no longer be neglected, and allowance must be made for the dependence of the shape and size of the chain on the concentration of the solution, which in turn depends on the nature of the solvent /195, 212/.

We must also remember that at finite concentrations of solutions, stable supermolecular formations arise, in which each macromolecule loses its individuality. Since no quantitative theory on the properties of concentrated solutions is available, we cannot develop a strict theory of the adsorption of macromolecules from solutions. Nevertheless, general quantitative relationships which describe the adsorption of polymeric molecules of different chemical nature from solvents of different power are interesting.

We studied /213/ the applicability of the Freundlich isotherm equation to the adsorption of polymers. This equation accurately describes the adsorption of low-molecular weight substances in the concentrational range where saturation of the adsorbent is observed. In these cases, the adsorption curve passes through a maximum, and the applicability of this equation was tested on the section before the maximum.

The results are given in Table 22, which shows the values of the coefficients β and μ in the Freundlich equation, determined by

$$x = \beta c^{\mu},\tag{6.1}$$

where x is the adsorption; c the equilibrium concentration of the solution; β and μ are constants. The same table gives the intrinsic viscosities of the solutions of the polymers studied that determine the solvent power.

Table 22 and Figure 83 show that the empirical Freundlich equation is applicable to the adsorption of polymers from solutions over a fairly wide concentrational range. In this case we can no longer speak of the existence of isolated macromolecular coils. Our research proves that this equation correctly describes adsorption also when secondary formations pass onto the surface instead of individual macromolecules.

An analysis of the data of Table 22 shows that factor β has a much higher value in the adsorption of polymers than in the adsorption of low-molecular weight compounds. It is usually close to unity, that is, adsorption is directly proportional to the concentration of the solution. Factor β, characterizing the adsorbability, is very dependent on the nature of the solvent. The selection of factor β as the magnitude characterizing adsorbability is much more convenient than saturation adsorption since the latter is frequently not observed.

The dependence of β on the nature of the solvent is related also to other factors connected with the appreciable differences between the adsorption of polymers and the adsorption of low-molecular weight substances. In the last case we deal with adsorbed particles of constant shape and composition, but in the adsorption of polymeric macromolecules the size and shape of the macromolecules continuously change with concentrational changes in the solution. Variations in the degree of aggregation and in the character of the supermolecular structures also occur.

Because of the complex dependence of the structure on concentration, temperature, and nature of the solvent, the adsorption of polymers from solutions cannot be governed by the same theoretical equations derived without allowance for the given state. These differences between the adsorption of polymers from solutions and the adsorption of low-molecular weight substances are very clearly indicated by the absence of a strict pattern of the variation in β with the nature of the solvent, characterized by $[\eta]$. This is completely understandable, since $[\eta]$ characterizes the size and shape of macromolecules in dilute solutions only. In concentrated solutions the more uncoiled shape of the chain, characteristic of better solvents, leads to the appearance of supermolecular structures. However, these will appear at higher concentrations than in poor solvents, because of the weaker interaction between the chains. Conversely, a very coiled chain in a poor solvent will also not lead to the appearance of aggregates, in spite of possible stronger molecular interaction. The optimum conditions for the appearance of structures in solutions will be reached when the chains are sufficiently uncoiled, and at the same time when they interact rather strongly. In other words, the solvent must be sufficiently poor for the supermolecular structures to appear at comparatively low concentrations.

All these factors will greatly affect adsorption involving transition of supermolecular structures onto the surface of the adsorbent. We shall consider, for example, the dependence of β on $[\eta]$ for poly (methyl methacrylate) solutions in different solvents (Figure 96). In this case

TABLE 22. Coefficients β and μ of the Freundlich equation in the range of applicability of the equation for the adsorption of polymers from different solvents on different adsorbents

System	t,° C	$[\eta]$	β	μ	Concentrational range where the equation is applicable, g/100 ml
On glass fibers					
Poly (methyl methacrylate)—acetone /38/	19	1.45	0.040	0.66	0.070—1.40
	25	1.55	0.166	0.94	0.050—0.80
	40	1.90	0.240	0.86	0.050—0.80
	60	—	0.251	0.74	0.050—0.80
Poly (methyl methacrylate)—acetone + 5% ethanol /38/	19	1.00	0.003	1.36	0.400—1.70
Poly (methyl methacrylate)—acetone + 5% heptane /38/	19	0.95	0.025	1.10	0.160—1.35*
Poly (methyl methacrylate)—toluene /38/	19	2.00	0.056	1.08	0.060—0.25**
	25	2.20	0.063	1.00	0.020—0.25**
Poly (methyl methacrylate)—chloroform /200/	19	4.40	0.025	1.00	0.17—1.27
	25	4.00	0.040	0.92	0.11—1.55
	40	3.40	0.056	1.02	0.10—1.30
Polystyrene—benzene /38/	25	1.40	0.028	0.93	0.05—1.50
	40	1.30	0.080	1.30	0.05—1.20
	60	1.20	0.251	1.07	0.05—1.30
Copolymer of styrene and methacrylic acid—benzene /200/	30	—	0.060	0.80	0.10—1.75
Gelatin—1 mole urea in water /200/	30	—	0.0006	1.03	0.0002—0.25**
Gelatin—water /200/	30	—	0.0001	0.41	0.01—0.17**
Polystyrene—cyclohexanone /199/	25	0.96	0	0	0.08—1.30
	40	0.70	0.063	0.91	0.08—1.30
	60	0.56	0.132	0.37	0.08—1.30
On iron powder					
Poly (vinyl acetate)—CCl$_4$ /77/	30.4	—	0.0022	0.226	0.001—0.02
	48.5	—	0.0028	0.225	0.001—0.02
	69.5	—	0.0031	0.225	0.010—0.02
On copper powder					
Poly (vinyl acetate)—benzene /173/	25	—	0.020	0.90	0.05—1.00
	35	—	0.045	1.15	0.05—1.00
	45	—	0.093	1.12	0.05—1.00
On aluminum powder /88/					
Poly (vinyl acetate)—dioxane	25	—	0.079	1.07	0.05—1.00
Poly (vinyl acetate)—dimethylformamide	25	—	0.051	1.00	0.05—1.00

* The equation cannot be applied at concentrations between 0.1 and 0.16 g/100 ml.

** The equation can be applied in the concentrational range up to the maximum of the isotherm.

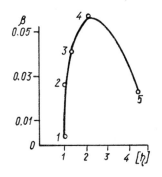

FIGURE 96. Dependence of β of the Freundlich equation on [η] at 19°C for poly (methyl methacrylate):

1) acetone—ethanol; 2) acetone—heptane; 3) acetone; 4) toluene; 5) chloroform.

the highest adsorbability (maximum value of β) will be observed at some medium solvent power, where evidently the conditions are most favorable for the formation of aggregates. The adsorbability is lower in very poor solvents (acetone + precipitating agents) and in a good solvent (chloroform). A temperature increase affects simultaneously the shape of the chain that changes the solvent power, and the variation in molecular interaction. Therefore, the dependence of β on [η] will change in character with variation in temperature.

For the system poly (methyl methacrylate)—acetone or toluene, that is, in poor solvents, [η] increases with temperature and β also becomes larger. This appears to be due to the more uncoiled shape of the chain and improvement in the conditions for the formation of aggregates. However, in good solvents, for example, in the systems poly (methyl methacrylate) − chloroform or polystyrene − benzene, [η] decreases with temperature, that is, some deterioration of the solvent is observed. However, β increases, which may be due to the enhanced molecular interaction.

We should note that for polystyrene in benzene, the second virial coefficient decreases with increase in temperature, which agrees with our data on the temperature dependence of [η].

Thus, β may both increase and decrease with [η] if we consider adsorption from moderately concentrated solutions. This is, as we have already noted, due to the simultaneous effect of two factors, namely, the shape of the chain and chain interaction during adsorption. Otherwise we would always observe maximum adsorption for coiled chains from poor solvents. Evidently, in poor solvents we will usually observe an increase in β with increase in [η], while in good solvents β will decrease with increase in [η]. Clearly, in all these cases we must take into account the difference in adsorbability of the solvent and its temperature dependence. However, adsorption of the solvent is usually negligible. It is usually disregarded even in the theoretical analysis of adsorption from dilute solutions.

There is thus an explicit relationship between the adsorbability and the nature of the solvent (this applies also to the heat of adsorption). However, there is no strict correlation between these magnitudes and the solvent power. Nevertheless, we clearly see the difference between the adsorption of low-molecular weight substances and the adsorption of polymers, in spite of the fact that both phenomena can be described by the Freundlich equation, and the connection between these features and the conditions of structure formation in polymer solutions.

Chapter 7

ADSORPTIONAL INTERACTION ON THE INTERFACE
AND THE PROPERTIES OF THE BOUNDARY LAYER

The adsorption of polymers on the interface with solids is very important in the reinforcing action of fillers, and in adhesion, gluing, etc. Adsorptional interaction is one of the important factors determining the properties of filled and reinforced polymers, the properties of the glue layers, the adhesion of polymers, etc. The fundamental patterns of adsorption processes considered in the previous chapters show that when a polymer is adsorbed on a solid surface the conformation of the macro-molecules changes; these conformations determine the structure of the adsorbed films, and the way this differs from the structure of polymers in solution or in the bulk. It is evident that many features of the structure of the film obtained by the adsorption of polymers by a solid surface from solution must be retained in systems in which adsorptional interaction between polymer and solid surface takes place in the absence of a solvent, that is, in all the practically important systems (reinforced and filled plastics, coatings, glues, etc.). It is very important to know the structure of the adsorbed films in such heterogeneous polymeric materials in order to understand the properties of the system, and to find ways of controlling them. However, adsorption methods with which it is possible to reveal some important features of the interaction between the polymers and solid surfaces and the behavior of polymers on the interface cannot give complete information on the properties of boundary layers in polymeric materials. This is because adsorptional interactions in solutions are not identical with interactions in the absence of a solvent, as the conformations of the macro-molecular chains in solutions differ from the conformations in highly elastic, vitreous, or crystalline and viscous-fluid states.

In this chapter we shall discuss the effect of adsorptional interaction on the interface on the properties of the boundary layers of polymers. We shall start with general ideas on adsorption.

BOUNDARY LAYERS OF POLYMERS
ON SOLID SURFACES

The molecular mobility of polymers in boundary layers is determined by the flexibility of the polymeric chain and the character of its interaction with the surface, that is, by the same factors that determine adsorption.

When we consider the problem of molecular mobility, we must bear in mind that the direct experimental determination of the molecular mobility in adsorbed polymer films is difficult. No papers have so far been published which describe research on adsorbed films.

We should note that it is necessary to define and separate concepts on the adsorbed film and the boundary layer. The adsorbed film is the layer of macromolecules formed on the surface as the result of the adsorption of a polymer from a solution, and in which some of the segments of the polymeric chains interact with this surface. The thickness of such an adsorbed film is determined by the conformation of the adsorbed molecules, but even in transition to more complex systems, in which polymolecular adsorption or adsorption of aggregates instead of molecules occurs, such a definition no longer holds, since in this case it is not only the molecules of the polymer in direct contact with the surface that are bound. The possibility was indicated by Silberberg /179—181/, and by Lipatov and Sergeeva /199—201/.

It is convenient here to introduce the concept of the boundary layer. We shall define it as a layer with properties that deviate from the bulk properties because of the effect of the surface. According to Rusanov /216/, the surface layer (or boundary layer) is characterized by the effective thickness. Beyond the limits of the effective thickness, the deviation in the local properties from the bulk properties is insignificant. The author shows that the introduction of this factor is possible because of the small radius of action of the intermolecular forces, which causes the rapid attenuation of the influence of one of the phases on some property of the neighboring phase. Such a definition corresponds to that of the Gibbs interface, to which the author ascribes properties of the inhomogeneous interphase region, where the effect of the mutual fields of forces of the two phases appears.

However, the concept of the effective thickness of the surface or boundary layer is very arbitrary when applied to polymers. This is due, firstly, to the chain structure of the polymeric molecules, so that the effect of the surface will spread over much larger distances from the surface than in the case of low-molecular weight substances. Secondly, the effective thickness of the intermediate polymer layer, with properties that differ from those of the polymer in the bulk, depends mainly on which property of the polymer is taken into consideration and whether this property is determined by the properties of the segments or of the macromolecules as independent kinetic units. Therefore, the different methods for determining the thickness of the boundary layer may give different values even for the same system.

This also explains why the type of function that describes the variation in the properties of the system with distance from the surface may differ. If we define the effective layer thickness, we also implicitly define the type of function that determines the change in the properties, that is, their non-uniformity at different distances from the interface. This problem is considered in some theoretical papers on the adsorbed film. They contain equations describing the distribution function of the macromolecules at different distances from the adsorbent surface, and also prove that such a distribution may differ in character.

When these definitions are applied to the boundary layers of polymers on solids, they must be considered with allowance for the conditions of

formation of the boundary layer. This is because in heterogeneous polymeric systems, where a knowledge of the boundary properties is of special
importance, the surface layers are formed by molding the material from
polymer melts in the presence of a solid surface (casting, extrusion,
pressing, etc.), or directly from oligomers during the hardening reaction
in the presence of the surface, and not by adsorption from solutions. The
conditions for adsorptional or adhesional interaction thus differ appreciably
from those during adsorption from solutions. We deal here with systems
concentrated to the limit, or monocomponent systems which do not contain
the solvent, where the conformation of the molecules does not correspond
to that in dilute solutions, and the forces of molecular interaction are
strong.

The conditions of formation of such a system preclude the direct study
of the properties of the boundary layer. In practically no case (except
when polymers crystallize in very thin layers) can the properties of the
boundary layers themselves be studied, and therefore all the conclusions
are based on the variations in the bulk properties of the polymers produced
by the interface, that is, on the detection of some excess characteristics.
Therefore, all the experimental characteristics represent the sum of the
properties of the boundary layer and the bulk, and conclusions on the
character of structural variations in the boundary layer are formed by
analyzing the direction of change of one or other characteristic. In this
case the most convenient model for studying the properties of the boundary
layers are filled polymers, which can be considered to be systems
consisting of solid particles coated by thin polymeric films.

EFFECT OF ADSORPTIONAL INTERACTION ON THE
MOLECULAR MOBILITY OF POLYMERIC CHAINS
IN BOUNDARY LAYERS

The adsorptional interaction of polymeric molecules with the surface,
that takes place in filled systems, may be considered to be a process
leading to redistribution of the intermolecular bonds in the system, and
the formation of additional nodes of the physical network structure as the
result of interaction between the segments and the surface. The additional
nodes formed should reduce the molecular mobility as the result of cross-
linking. We should expect that the number of additional nodes and the
properties of the surface layer of the polymer will vary, depending on the
conditions of formation of the filled polymer, and the type of chain interaction with the surface. The first act of formation of the surface film
(lacquer coats, glued joint) is the adsorption of polymeric molecules by the
surface. The properties of the boundary layers will vary, depending on the
character of the adsorption and the shape of the chains in the melt or
solution.

Studies on relaxation processes in polymers, that occur on the interface
with solids, are of practical and theoretical interest in connection with the
creation of filled polymers and the finding of the optimum conditions for
their processing and application.

The methods of dielectric relaxation and NMR are at present the most popular for studying the molecular mobility of polymers. We were the first to use these methods for determining the variations in the molecular mobility of chains in filled polymers /217—219/. We shall not discuss these results in detail (some are generalized in monograph /18/), but show the main trends in the variations in the molecular mobility in boundary layers.

We found that the presence of an interface leads to an essential change in the relaxational behavior of a polymer in the boundary layer, changes in the glass point of the polymer and width of the glass transition range, changes in the mean relaxation period, etc. This is the result of variations in the molecular packing density and decrease in the mobility of the polymeric chain segments and of larger kinetic elements because of their interaction with the solid surface.

For our studies we used poly (methyl methacrylate), polystyrene, the copolymer of methyl methacrylate and styrene, different polyurethan elastomers, and cellulose acetate. These polymers differed in the presence of functional groups in chains, in their potential ability to interact with the solid surface, and in the flexibility of the molecular chain. The boundary layers of the polymers were modeled by introducing finely dispersed fillers in different quantities into the polymer. It was thus possible to consider that such a filled polymer consisted of solid particles coated by a polymer film varying in thickness. Depending on the amount of filler used, the boundary layers produced had a thickness that varied between 0.5 and 5 μ in our experiments. We used Aerosil silica as a body with high surface energy and polytetrafluoroethylene (teflon) as a body with low surface energy.

The molecular mobility was investigated by the impulse NMR method, and by studying the dielectric relaxation.

We shall now discuss the effect of the chemical character of the interface on changes in the molecular mobility of polymeric chains in boundary layers. We shall characterize these variations by the displacement of the maximum of the dielectric losses on the $\tan \delta = f(T)$ curve. As the thickness of the boundary layer decreases, the maximum $\tan \delta$, corresponding to the dipole-group relaxation process, is displaced toward lower temperatures, and the maximum of the dipole-segment relaxation process toward higher temperatures. This indicates changes in the mean relaxation times of the corresponding processes in the boundary layers.

It was shown that for comparable thicknesses of the boundary layers in surfaces with high and low surface energies, both maxima are displaced by the same amount. A similar pattern was observed when the temperature dependence of the proton spin-lattice relaxation was studied by the impulse NMR method. This proves that the effects observed in studies on dielectric relaxation are not the result of the Maxwell — Wagner effect of the inhomogeneity of the medium, characteristic of objects with conducting and nonconducting regions /223/.

From these data it is possible to calculate the temperature dependence of the parameter α of the distribution of the relaxation time in the boundary layer by the method of the Cole-Cole circular plot /224/. It was found that in the boundary layer the above parameter decreases. This corresponds to a widening of the spectra of the relaxation time, and agrees with the divergence of the maxima of the dipole-group and dipole-segment losses.

It is interesting to note that the variation in the magnitude of the maxima of the losses or of the times of spin-lattice relaxation is independent of the nature of the surface. Such an independence is also observed in studies on molecular motion in the boundary layers caused by the mobility of larger structural units, which can be characterized by the mean relaxation time of the processes of isothermal reduction in volume /218, 220, 221/.

We also studied the dependence of the dielectric and spin-lattice relaxation in the bulk and on the surface of Aerosil silica particles, when the surface was unmodified, or modified by dimethyldichlorosilane. This modification leads to a considerable change in the surface energy of the solid particles. The tests were carried out with polyurethan rubbers of different chemical nature. These had a much higher chain flexibility than polystyrene or poly(methyl methacrylate). For the given class of polymers, the general pattern of the dispersion regions remained unchanged. The variation in the molecular mobility of the chains on the interface was practically the same for the modified and unmodified surfaces. This is in agreement with the hypothesis that these effects are independent of the chemical nature of the surface, although in this case the polymers had active functional groups in the molecular chains that could form hydrogen bonds with the surface of the Aerosil silica.

From these data we can conclude that the change in the molecular mobility of the chains in the boundary layers does not involve variations in the phase interaction energy only. It is also interesting to note that the observed change in the molecular mobility applies not only to the boundary layers in direct contact with the surface (in this case we would not have observed macroscopic effects), but also to large distances from the interface, although the magnitude of the effect is nonlinearly dependent on the thickness of the boundary layer.

To come to some conclusion on the reasons for the observed variations in the molecular mobility we shall examine the data for the boundary layers of a stiff-chain polymer, namely, cellulose acetate. The $\tan \delta = f(T)$ curve for cellulose acetate in the bulk and on the surface of modified and unmodified Aerosil silica shows that in the case of a stiff-chain polymer, no changes in the molecular mobility occur near the interface.

From this we can conclude that the main role in the changes in the molecular interaction of polymeric chains in the boundary layers is played not by energetic chain-surface interaction (which cannot propagate far from the surface onto a layer that is not in direct contact with it), but by conformational changes in the polymeric chains near the interface.

We shall start from the theoretical concepts on the conformation of chains of polymeric molecules near the interface /225, 226/. The distribution function of the end-to-end distance of the polymeric chains near the interface differs from that in the bulk. The solid surface is a reflecting barrier, which does not allow the macromolecules to take up the same number of possible conformations as a macromolecule in the bulk. There is thus a reduction in the number of conformations, or, in other words, the chain becomes stiffer. We believe that this is the main reason for the changes in the total relaxation behavior of macromolecules in the boundary layer.

We therefore think that the limitations on the mobility of the chains in the boundary layers are related to the entropy factor, that is, reduction in the number of conformations which may be taken up by the macromolecules near the interface. It is thus possible to explain satisfactorily the independence of the effect of the chemical nature of the interface, the propagation of the change in the mobility to layers not directly in contact with the surface, and finally the effect of the flexibility of the polymeric chain on these factors. Naturally, the number of conformations of the molecules of a stiff-chain polymer, which is much smaller than for flexible molecules, cannot vary as much near the interface as in the case of flexible molecules, because of the stiffness of the chain. Here, the effects of changes in the mobility of the chain are not observed.

We can thus conclude that variations in the molecular mobility are due to a decrease in the chain flexibility in the boundary layer because of conformational restrictions imposed by the geometry of the surface. It follows from our data that it does not matter whether the conformational changes are due only to the presence of the surface or to some degree of binding of the molecules by the surface. The last factor is very important from the point of view of the strength of the adhesion bond, but is of no importance for the decrease in the molecular mobility, since these processes do not involve impairment of the bonds on the interface.

In all the examples given we did not consider the case of strong specific interactions on the interface, where the pattern will possibly differ somewhat from the one described.

It would therefore be interesting to estimate the contribution of the energetic and entropic factors to the change in molecular mobility near the interface. We made such an attempt /260/ using data on the activation energy of the relaxation processes in the boundary layers determined from the temperature dependence of the mean relaxation times (Table 23). We started from the expression

$$\tau = \tau_0 \exp{(\Delta F / RT)}, \tag{7.1}$$

where ΔF is the free energy of activation of the relaxation process; τ is the relaxation time of the processes; τ_0 is the value τ at $1/T = 0$. From this equation we obtain

$$\ln{(\tau/\tau_0)} = \Delta F / RT, \tag{7.2}$$

$$\frac{\partial \ln{(\tau/\tau_0)}}{\partial (1/T)} = \frac{\Delta F}{R} + \frac{1}{RT}\frac{\partial \Delta F}{\partial (1/T)} =$$

$$= \frac{\Delta F}{R} - \frac{T}{R}\frac{\partial \Delta F}{\partial T} = \frac{\Delta F}{R} - \frac{T\Delta S}{R} = \frac{\Delta H}{R} \tag{7.3}$$

or

$$\Delta H = R\frac{\partial \ln{(\tau/\tau_0)}}{\partial (1/T)} = -RT^2\frac{\partial \ln \tau}{\partial T}, \tag{7.4}$$

where ΔH is the activation enthalpy when τ_0 is independent of T. Hence

$$TΔS = -ΔF + ΔH = -RT \ln(τ/τ_0) - RT^2 \frac{\partial \ln τ}{\partial τ}, \qquad (7.5)$$

$$ΔS = -\frac{\partial [RT \ln(τ/τ_0)]}{\partial T}. \qquad (7.6)$$

TABLE 23. Activation energy and temperature shift of relaxation processes of polymers applied in thin layers, determined by the NMR and dielectric methods

Content of Aerosil silica, %	Content of teflon, %	Activation energy of relaxation, kcal/mole			Activation energy of dielectric relaxation, kcal/mole		
		PMMA	PST	copolymer of MMA and ST	PMMA	PST	copolymer of MMA and ST
				Group movement			
0	0	1.8	—	2.1	23.7	—	14.9
8.83	—	—	—	1.7	—	—	12.6
1.32	—	1.4	—	—	18.5	—	—
23.08	—	1.2	—	—	15.4	—	—
24.90	—	—	—	1.8	—	—	10.7
—	26.5	—	—	1.7	—	—	12.6
—	49.2	1.5	—	—	20.0	—	—
—	75.0	1.4	—	1.5	18.8	—	10.6
				Segment movement			
0	0	14.5	11.3	13.3	—	90.0	99.0
8.83	—	—	—	12.0	—	—	—
1.32	—	9.8	—	—	—	60.9	—
23.08	—	9.2	12.3	—	—	57.1	—
24.90	—	—	—	11.5	—	—	85.5
—	26.5	—	—	12.0	—	—	89.5
—	49.2	11.0	—	—	—	69.2	—
—	75.0	10.1	13.1	11.4	—	63.2	84.6

From the experimental dependences it is possible to determine the thermodynamic characteristics of the activation process.

The change in the activation entropy in the boundary layers is much larger than in the bulk, while the entropy decreases very slightly. These results also show that the change in the molecular mobility near the interface is due mainly to conformational effects. The corresponding values for the other systems that we studied are given in Table 24, which shows that with decrease in the thickness of the boundary layer, the change in the activation energy increases appreciably. This increase is quite understandable, if we take into account that the transition into a new position under the action of a field requires a much greater conformational change in a stiff molecule than in a more flexible one, although the height of the energetic transition barrier may remain the same.

TABLE 24. Enthalpy and entropy of activation of segmental processes in the boundary layers of polymers

Polymer	Content of Aerosil silica or teflon, %	ΔH, kcal/mole	ΔS, cal/mole·deg
Copolymer of MMA and ST	0	13.3	1.10
	8.83 a	12.0	1.52
	24.90 a	11.5	1.81
	26.50 t	12.0	1.50
	75.00 t	11.4	1.82
Polyester-based PU	0	2.0	0.13
	13.30 a	1.8	0.20
	13.30 a	1.8	0.21
Polyether-based PU	0	4.4	1.15
	14.80 a	3.8	1.79

Note. "a" and "t" denote Aerosil silica and teflon, respectively.

These data indicate the considerable influence of the interface with a solid body on the mobility of molecular polymeric chains present on this interface, and their dependence on the boundary layer thickness. The widening of the relaxation time spectra observed indicates that the interface exerts different influences on the mobility of the individual relaxation agents taking part in the overall movement. The main reason for the changes in the relaxation behavior of the polymeric chains is reduction in the number of possible conformations at the interface as the result of conformational restrictions imposed by the surface, or because of interaction with the surface.

It can be seen that the results more or less agree with the theoretical concepts on adsorption, and in particular with the concepts on the important role of conformational changes in the boundary layer, which take place even when the interaction between the polymeric molecules and the surface is weak (see Silberberg's theory). The increase in the molecular mobility of the chain sections not bound to the surface can be explained by the existence of adsorption loops on the surface, which are not directly bound to the surface and hence have a higher molecular mobility.

Another fact is worth noting. The change in the molecular mobility in the boundary layers should not be considered to be the result of adsorptional interaction caused by change in the enthalpy of the system only. In principle, the same results can be obtained for systems with strong and weak interactions of the chains with the surface, where all effects of change in the molecular mobility are the result of entropy factors. Hence, changes in the mobility cannot be used as a characteristic of the adhesion of the polymer to the surface.

Later research on the molecular mobility in filled systems confirmed the fundamental assumptions developed in our papers /227, 228/.

CHANGES IN THE PROPERTIES OF BOUNDARY
LAYERS AS THE RESULT OF A DECREASE
IN MOLECULAR MOBILITY

The limitations on molecular mobility as the result of adsorptional
interaction lead to appreciable changes in the properties of the boundary
layers of polymers. They appear in the packing density of the molecules
in the boundary layers, in the glass point and relaxation behavior of filled
polymers, and in the character of the supermolecular structures formed
on the surface.

Packing density in boundary layers

We were the first to find a loosening of the molecular packing in the
boundary layers of polymers /229—231/. We shall discuss some thermodynamic data.

From studies on the adsorption of vapors by polymers it is possible to
calculate the change in the thermodynamic functions during sorption. In
the theory of polymeric solutions this process is usually considered to be
the mixing of the polymer with the solvent, determined by the energies of
segment — solvent and segment — segment interactions, and by the flexibility
of the chains that affect the entropy of mixing.

A decrease in the thickness of the boundary layers leads to a large
increase in sorption. Calculation showed that this increase cannot be
caused by the sorption of vapors on the surface of the solid, but by
structural changes only. If we use a solvent for the given polymer as
sorbent, then we can calculate from the usual thermodynamic relationships
the change in the partial free energy during sorption (but under the condition
that the system is athermal), and the change in the partial specific entropy
of the polymer in the bulk and on the surface.

When we studied the sorption of ethylbenzene vapors by polystyrene with
various amounts of glass fibers, we found that ΔS_2 increases with increase
in the filler concentration in the polymer film. In terms of classical
concepts on the theory of solutions, this means that the polymeric molecules
are arranged in more ways in the boundary layer than in the bulk. The
increase in sorption also indicates a loosening of the packing of the
macromolecules in the boundary layers.

According to the theory of solutions, swelling is an important characteristic of polymers. In terms of the Flory theory, swelling is determined
by the number of junctions in the three-dimensional network of the polymers,
and can be used for their determination.

When we studied the dependence of the degree of swelling on the polystyrene content on the surface of glass fibers, we found /232/ that with
increase in the thickness of the polymeric film on the fibers, swelling
continually decreases, and only when the polymer content is about 200%
of the weight of the fibers, is the swelling close to that in the bulk. These
data not only confirm that there is a loosening of the packing of the

molecules on the surface, but they also indicate the greater distance from the surface at which its effect can still be felt.

In the above case there is no strong interaction between the polymer and the surface. If there is a strong interaction, the pattern may differ appreciably. The dependence of the effective density of the network of three-dimensional polyurethans applied to a solid surface on the thickness of the coating was studied /240/. In this case additional bonds are formed with the surface, and lead to an increase in the density of the network. With increase in the thickness of the layer, this effect weakens, and becomes insignificant at a distance of 200 μ from the surface. Hence, in the case of this polymer the effect of the surface can be felt over a larger distance from the surface.

Thermodynamic studies thus indicate considerable differences in the structure and properties of the boundary layers. Similar results were later obtained in many papers. The thermodynamics of swelling and of the sorption of filled polymers are described in detail in /18/.

We explain the loosening of the packing in the boundary layer as follows. The formation of adsorptional bonds with the surface during the molding of the polymeric material leads to additional cross-linking of the system, and this considerably restricts the mobility of the polymeric chains near the surface. This causes a change in the conditions of the course of the relaxation processes, and hinders the establishment of the equilibrium state near the surface. Hence, the appearance of a closely packed structure is impossible under such conditions. The effect of the conditions of the relaxation process on the packing density of polymers is discussed in /233/.

We also assume that the process of the formation of supermolecular structures on the surface occurs partially and spontaneously. We can suppose that for the same reasons the molecular aggregates or other supermolecular structures may be less closely packed. With increase in the surface of the filler, the chain mobility becomes more restricted, even during the formation of the boundary layer, and the packing of the macromolecules in it becomes looser. When the formation of the material is complete, and the aggregates and the molecules are more loosely packed and bound to the surface, restriction of the mobility of the molecules in the boundary layer has the greatest influence on the properties.

Glass point of the boundary layers

Changes in molecular mobility lead to essential changes in the glass point of amorphous linear and network polymers in the boundary layers. These changes are confirmed by the data reported on the molecular mobility and by numerous measurements that were carried out by us and other research workers of the glass point in boundary layers. A thermodynamic inter-pretation of the change in the glass point of polymers on the interface with the solid phase is given in the paper of Lipatov and Privalko /234/.

It is known that the transition from a highly elastic to a glass state is a cooperative process, and therefore the jump in specific heat on glass transition evidently depends on the number of molecules or segments taking part in this transition. Since transition involves mobility of the macromolecules, an increase in the jump in specific heat may be uniquely related to the exclusion of some macromolecules from participation in the process. Experimental data confirm this hypothesis: with increase in the content of the solid phase, the jump in the specific heat decreases. It is thus possible to estimate the fraction of the polymer present in the boundary layers. If we assume that the macromolecules present in the boundary layers near the surface do not participate in the overall process, then the fraction of the "excluded" macromolecules is

$$v = (1 - f) = 1 - \Delta C / \Delta C_a, \qquad (7.7)$$

where ΔC_a and ΔC are the jumps in specific heat for unfilled and filled samples, respectively. From these data we can determine the thickness of the boundary layers as follows. If, for the sake of simplicity, we consider the filler particles to be spheres of radius r and denote the thickness of the adsorbed film by Δr, then the volume of the adsorbed film round the filler particles will be described by

$$V = \frac{4}{3} \pi [(2 + \Delta r)^3 - r^3]. \qquad (7.8)$$

However, the volume fraction of the boundary macromolecules can be represented as $(1 - f) c$, where f is the fraction of the nonbound macromolecules, and c is the total volume fraction of the polymer in the system. By equating the ratio between the volume of the adsorbed film round the particle and the particle volume to the ratio between the volume fraction of the boundary macromolecules and the volume fraction of the filler in the system, we can write

$$\left(\frac{r + \Delta r}{r} \right)^3 - 1 = (1 - f) \frac{c}{1-c}. \qquad (7.9)$$

If we take the experimental data for the system oligoethylene glycol adipate − Aerosil silica $(1 - f) \cong 0.1$ and $c = 0.975$, then $\frac{\Delta r}{r} \cong 0.8$. Since the Aerosil silica particles have a diameter of 250 Å, the thickness of the adsorbed film is 100 Å, We obtained similar magnitudes of the order of 170 Å for linear polyurethans filled with carbon black.

Thus, the absolute value of the specific heat of the polymeric phase in filled systems is lower than in unfilled ones. This is because the chemical potential in the boundary region is lower than that in the bulk. Thus, the thermodynamic data indicate certain structural changes in the boundary layers near the polymer − solid body interface.

We have already mentioned that the boundary layer thickness depends on the properties of the solid surface and on the characteristics of the polymeric phase. The effect of the chemical nature of the polymer on the changes in the properties of the boundary layers is very appreciable. We shall examine some published data obtained by measuring the specific heat (Table 25). This table shows that with increase in the content of

Aerosil silica in the polymers, the jump in the specific heat ΔC_p at the glass point always decreases more or less sharply. This indicates the transition of some of the macromolecules from the bulk into the boundary layer near the solid surface. Table 25 also shows the fraction ν of the polymer found from (7.7). The value of ν increases with increase in the content of filler in the system (although no proportionality is observed), and tends to some limit. We attempted to analyze our data according to the theory of the free volume, which we used in earlier research on glass transition in filled systems studied by means of variations in specific volume /220, 221/.

TABLE 25. Glass transition parameters in filled polymers

Content of Aerosil silica, wt %	T_c, °C	ΔC_p, cal/mole	ν	E_0, cal/mole	ν_c, cm³/mole	ε_h, cal/mole	V_h, cm³/mole
			Polystyrene				
0	95	6.25	—	7320	100.5	1230	16.9
1	95	5.60	0.105	—	—	1375	18.9
5	95	4.55	0.270	—	—	1705	23.5
10	95	3.10	0.505	—	—	2160	29.7
15	95	3.00	0.520	—	—	2190	30.1
			Poly (methyl methacrylate)				
0	105	10.00	—	11,380	85.9	1180	8.9
1	110	9.80	0.020	—	—	1215	9.2
5	118	9.00	0.100	—	—	1350	10.2
7	121	8.40	0.160	—	—	1455	11.0
10	123	8.10	0.190	—	—	1530	11.5
			Polyurethan				
0	-34	19.60	—	16,380	143.0	895	7.8
1	-33	17.20	0.120	—	—	1020	8.9
5	-32	15.80	0.195	—	—	1115	9.7
10	-30	14.60	0.255	—	—	1170	10.2
20	-30	14.20	0.275	—	—	1200	10.5
			Polydimethylsiloxane				
0	-125	7.20	—	4985	65.0	695	9.0
10	-124	6.50	0.095	—	—	755	9.8
30	-123	5.82	0.190	—	—	805	10.5
50	-123	5.33	0.260	—	—	845	10.9

According to the theory of Wunderlich /235/, ΔC_p is given by

$$\Delta C_p = \frac{E_c}{T_c} \cdot \frac{\varepsilon_h}{RT_c} \exp\left(-\varepsilon_h/RT_c\right). \qquad (7.10)$$

where T_c is the glass point; E_c is the molecular cohesion of the polymer at the glass point; ε_h is the energy of "hole" formation.
The molar hole volume is determined by

$$V_h = (\varepsilon_h V_c)/E_c, \qquad (7.11)$$

where V_c is the molar volume at T_c. Table 25 gives the values of ε_h and V_h calculated from the above equations. For unfilled polymers, V_c was determined by extrapolating specific volume curves to T_c. The values of E_c were calculated by Wunderlich's method. The value of ε_h should depend on the overall contribution of intra- and intermolecular actions. The main contribution to ε_h is given by intramolecular action, because ε_h decreases continuously from high values for polystyrene and poly (methyl methacrylate to lower ones for polyurethan and polydimethylsiloxane, which have flexible chains.

From these ideas on the change in the structure of the boundary layers we can conclude that the values of ε_h and V_h for these layers will differ from those in the bulk of the polymer. If we assume, to a first approximation, that E_c and V_c for filled systems differ only slightly from the values for unfilled ones, we can calculate ΔC_p and T_c from the experimental values of ε_h and V_h (see Table 25).

We see that the introduction of fillers into the polymers leads to a considerable increase in ε_h and V_h. We shall now explain these results. The increase in ε_h with increase in the content of the solid phase in the system indicates an increase in the energy required for segmental mobility during transition of a glass state into a highly elastic one, and this in turn indicates the restricted mobility of the macromolecules in the boundary layer. However, an increase in V_h is related to the fact that the polymer chains in the boundary layer are more loosely packed than those in the bulk. Thus, calculations lead to the conclusions that were earlier formed on the basis of a quantitative analysis of the changes in the properties of the system.

From these estimates it is possible to consider the dependence of the properties of the polymer in the boundary layer on the chemical structure of the chain. It was proved that for filled systems the glass point and the ratio $V_h/V_{h,0}$ increase in proportion to ν:

$$T_c = T_{c,0} + \Delta T \nu, \qquad (7.12)$$

$$T_h/V_{h,0} = A + B\nu. \qquad (7.13)$$

In these equations ΔT is the maximum increment in the glass point for a system in which the whole polymeric phase is subjected to the influence of the surface ($\nu = 1$). The values of ΔT correlate well with the hole density of the cohesion energy ε_h/V_h (Table 26). Thus, we can conclude that the greatest changes in the properties of the boundary layers are observed in systems with large values of ε_h/V_h, indicating the intensity of intermolecular action in the bulk of the given polymer. This magnitude determines the character of the dependence of the glass point on the polymer fraction in the boundary layer. By analogy, coefficient B represents the maximum increment of the free hole volume in a system with $\nu = 1$.

TABLE 26. The values of the coefficients in (7.12) and (7.13)

Polymer	ΔT, °K	ε_h/V_h, cal/cm³	A	B
Polystyrene	0	72.8	0.95	1.72
Poly (methyl methacrylate)	105	132.0	1.00	1.47
Polyurethan	14	114.0	1.00	1.17
Polydimethylsiloxane	7	77.0	1.00	0.83

It can thus be concluded that the properties of filled systems are determined by the fraction of the polymer present in the boundary layer.

In connection with theoretical ideas on adsorption, we should mention the results that we obtained when we studied the glass point of plasticized filled polymers /221, 236/. We found that for the same content of plasticizer the glass point of the filled polymer decreases more sharply than that of the unfilled polymer. When the content of plasticizer increases above a certain limit, the glass point of filled films becomes lower than that of unfilled films. These data indicate that there is competition between polymer and plasticizer for sites on the surface, and that the polymer is displaced by surface plasticizer molecules. This agrees with the ideas on the adsorption of polymer mixtures. We should note that the papers of Lipatov and Geller /220, 221/ on volume relaxation in filled polymers proved that the relaxation time and the activation energy are little dependent on the nature of the surface for a given polymer. This confirms the conclusion that in the restriction mechanism of the mobility of the chains near the interface, the largest role is played by processes that involve a decrease in the possible number of configurations (determined by the flexibility of the macromolecules), and not by energetic interaction between polymer and surface.

STRUCTURE OF BOUNDARY LAYERS OF CRYSTALLINE AND CROSS-LINKED POLYMERS

Most experimental data on changes in the properties of the boundary layers of polymers on a solid surface relate to linear polymers, because changes in the properties produced by introducing an interface can be fairly easily observed for these polymers. These studies are much more complicated for crystalline and three-dimensional polymers. Lipatov and Geller /220, 221/ found that with increase in the density of the three-dimensional network of the polymer, the effect of the surface on the glass point becomes weaker. This is because the chains become more rigid when the network density increases. For rigid polymers of the acetylcellulose type, the solid surface also does not affect the transition temperature /219/. This, however, does not mean that in the presence of a solid surface in systems no structural changes in the boundary layers take place. To study the changes in the structure and properties of the boundary layers in crystalline or cross-linked polymers, we need new physical methods for investigating the boundary layers.

In many of our papers, and in the papers of other authors (especially that of Malinskii /69/) it was found that the presence of an interface has an appreciable effect on the crystallization of polymers and the morphology of crystalline formations. Thermodynamic studies on the crystallization processes of polymers (poly(ethylene glycol adipate)) in the presence of a filler showed that with increase in the content of filler the relative fraction of the crystalline region decreases slightly, while the number of macromolecules with a lower mobility increases. In other words, the structure of the amorphous regions of the crystalline polymer changes, and not the crystalline structure itself. This fact complicates the use of X-ray diffraction methods in the study of the boundary layers of crystalline polymers.

We therefore studied the time of spin-lattice relaxation T_1 of radicals stabilized in the bulk and in the surface layer of a crystalline polymer. By means of relaxation studies it is possible to conjecture about the mutual interaction of radicals which are affected by their stabilizing matrix /237/.

The radicals in the bulk of the polymer were generated by γ-irradiation, and in the surface layers they were generated by the action of an electrodeless high-frequency discharge.

Since treatment with high-frequency discharge affects only the surface layers, the generated radicals are stabilized in those layers which have a structure that is considered to be different from that in the bulk. The thickness of the layer in which the radicals are stabilized is 0.1 to 1 μ.

The polymers studied were polycaprolactam fiber and polyurethan prepared from hexamethylene diisocyanate and butane-1,4-diol. The samples were irradiated by ^{60}Co at a pressure of 10^{-3} mm Hg and a temperature of 77°K. The intensity of irradiation was 400 rad/sec, and the total dose was 5–6 Mrad. EPR spectra showed that the radicals generated by these two methods had the same chemical nature.

When the data on the effect of the matrix structure on T_1 are analyzed, we must allow for the concentrational dependence of this parameter. We found that at irradiation doses of 6 to 12 Mrad, T_1 is slightly dependent on concentration. This was also used as the criterion for selecting the irradiation dose, as such a dose does not lead to cross-linking. The time of action of the high-frequency discharge was selected so that T_1 was concentration independent.

Dipole spin-spin interaction, exchange interaction, and spin-orbital interaction, are the main mechanisms of spin-lattice relaxation. In our case the first two types of interaction do not contribute to T_1. Therefore, the change in this parameter is due to spin-orbital and hyperfine interactions, which are determined by intra- and intermolecular motions in the polymer matrix. Thus, the mobility of the matrix may affect T_1.

Published data and our own studies on the concentrational dependence of the spin-lattice relaxation time (T_1 decreases with increase in the local concentration of the radicals) showed that a considerable difference between the values of T_1 for radicals stabilized in the surface layer and in the bulk may indicate that the radicals in the surface layer are more closely arranged. The average limiting concentrations calculated per gram of polymer are similar for the two types of action, but when high-frequency discharge is employed, the free radicals are accumulated in a much smaller volume. Therefore, the local concentrations of the radicals in the surface layer are much higher than the local concentrations of the radicals stabilized in the bulk by irradiation.

The higher concentration of the radicals in the surface layer can be explained by the higher defectiveness of this layer. Radical accumulation as the result of the high-frequency discharge takes place by migration of the radicals within the surface layer toward its structural defects. Hence, a higher local concentration may lead to a decrease in T_1. However, T_1 is not determined by the radical concentration alone. A low value of T_1 may also be due to high intramolecular mobility near the radical centers in the macromolecules because of a decrease in the orderliness of the surface layer.

We can thus consider that our experimental data indicate more defects in the surface layer than in the bulk.

We shall now consider the boundary layers of network polymers. We studied the structure of the boundary layers of hardened polymer binders in fiber glass, the effect of the nature of the binder on the density of the boundary layer, and also the conditions of thermal treatment after hardening. The structural changes were studied by the molecular probe method /238, 239/. The method is based on a study of the luminescence spectrum of anthracene impurity molecules used as the probe. The matrix structure affects the luminescence spectrum, and from the position of the spectra of anthracene molecules introduced into the system the authors determined the density of the surroundings (matrix) and its changes due to structural influences.

The samples were prepared from epoxy resin ED-5 with various active and inactive plasticizers. The systems used, which are cold-hardened systems, were heat-treated under different conditions. Anthracene was introduced in various ways: a) via an intermediate solvent (xylene) into the nonhardened epoxy resin, with subsequent removal of the solvent before hardening; b) into the hardened resin by swelling the samples in a solution of anthracene in xylene, followed by heat treatment; c) into the heat-treated samples. Thus, we could estimate in case (a) the average density of the polymer, since the molecular probes are uniformly distributed over the sample. In case (b) the effect of the solvent on the polymer with and without filler and the effect of later heat treatment could be estimated. In case (c) the appearance of heterogeneities in the sample as the result of heat treatment, when the molecules of the probe penetrate into the most loosened regions of the samples after swelling in xylene, could be estimated. Therefore, in the first case we determined the mean density, and in the second case the loosening of the sections.

It was found that polymeric systems without a filler may become both denser and looser as the result of heat treatment (the density changes by 0.5 to 2%), depending on the nature of the plasticizer. When a filler is introduced into composites, the mean density of the hardened polymer does not change, or increases slightly (by not more than 0.5%). However, all the samples become structurally inhomogeneous, so that in the samples we observe regions with densities that are 5–6% lower than those in the initial material. Heat treatment at 80°C has practically no effect on the mean densities, but makes the structure more uniform: the density of the loosened regions increases by 4—5%, apparently because of stress relaxation.

Further heat treatment at 160°C again leads to heterogeneity, to a somewhat lower average density, and to the appearance of regions of lower density. These data indicate that the introduction of a polymer-filled (solid) interface leads to structural changes in the three-dimensional polymer, and to the appearance of loosened stressed regions. Thus, in filled polymeric systems there are loosened regions with a density that is 5—6% lower than that in unfilled systems. However, the retarding effect of the filler on structure formation may be decreased or even reduced to zero by subsequent heat treatment.

The heat treatment of filled polymeric systems thus has a decisive influence on their structure, and partially or completely eliminates structural differences produced by the introduction of a filler into the system. It is evident that the bonds between the polymeric molecules and the surface are redistributed. These data confirm the correctness of the general concepts on the behavior of the boundary layers of polymers at the interface that we have already proposed, and make it possible to extend these concepts to many heterogeneous polymeric materials. The data show that it is possible to influence the properties of polymers by varying the conditions of production.

These changes in the structure of a filled three-dimensional polymer in various degrees of curing (since heat treatment leads to further curing of the system) can be followed in various stages of the process by estimating the molecular mobility of the growing polymeric chain by measuring the dielectric relaxation /261/.

It was found that during the process of curing the solid surface, that restricts the mobility of the polymeric chains, the kinetics of the process of binding the molecules into the three-dimensional network are affected. The curing process takes place more slowly in filled samples, that is, the relaxation processes of the chain segments take place at higher temperatures. The last results indicate that changes in the molecular mobility of the chains during the formation of the three-dimensional network appreciably affect the character of the reaction and the reaction rate, and also the final structure of the three-dimensional polymer (this can be seen from the luminescence spectra). The corresponding structural changes are related to changes in the relaxation behavior of the polymers near the interface.

The general pattern of the behavior of macromolecules near the interface with a solid body that we elaborated for linear amorphous polymers is thus also correct for three-dimensional and crystallizing polymers. It should be borne in mind that the very existence of an interface is important. Naturally, in a crystalline polymer there is no interface with a solid body, but there is an interface with air. However, the effect of this interface is distinct, and we can no longer speak of interaction with the surface, since all the changes are due to the action of entropy factors only.

EFFECT OF A SOLID SURFACE ON THE
SUPERMOLECULAR STRUCTURE OF POLYMERS

Adsorptional interaction on the polymer — solid body interface affects the conditions of formation of the polymeric material and leads to changes in the supermolecular structure of the boundary layers and of the total polymeric phase in the filled system. Kargin and Sogolova /256/ showed that the introduction of solid additives into crystallizing polymers makes it possible to control the size and number of spherolites. The mechanism of the action of additives is that ordered regions of the polymer that function as crystallization centers are formed on the surface of the solid particles. However, Malinskii /257, 258/ found that polymer crystallization near the walls is inhibited by a solid surface.

Kuksin et al. /240/ studied the effect of a solid surface on the super- molecular structures in cross-linked polymers (polyurethans) used as coatings for various solid surfaces. The authors found that the character of the supermolecular structures is determined by the type of surface, and depends on the density of the three-dimensional network. These authors were the first to make a layer-by-layer analysis of the super- molecular structures at various distances from the surface. It was found that the morphology changes with increase in the distance from the surface, and fine globular closely packed structures pass into large globular structures with globule aggregation. The effect of the surface on the supermolecular structure extends over large distances from the surface. It is only at a distance above 160μ that the structure of the films formed on a solid surface becomes similar to that of the film formed on the polymer — air interface.

RELATIONSHIP BETWEEN THE ADSORPTION OF
POLYMERS AND THE ADHESION OF POLYMERS
TO A SURFACE

Adsorptional interaction on the interface is determined by the adhesion of the polymers to the solid surface. In the thermodynamic meaning, adhesion is the work required to overcome the forces of cohesion of two different surfaces. From this point of view a certain relationship must exist between polymer adsorption and polymer adhesion on a solid surface. Unfortunately, although in many papers on adsorption the problem of such a relationship has been considered, in practice this problem has never been solved, that is, there are no data on a comparison of adsorption with the thermodynamic work of adhesion. This may be because the estimation of the thermodynamic work of the adhesion of a polymer to a solid surface is very complicated. Therefore, adhesion is usually characterized by the nonequilibrium work of breaking away and not by the thermodynamic equilibrium magnitude.

However, breaking away practically never occurs between two materials, as shown by Bikerman /12/. All destruction of an adhesional connection includes cohesional destruction; truly adhesional destruction is very rare, and even in this case breaking away is irregular. Therefore, a direct comparison between adhesional characteristics found in various papers and data on the adsorption of the same polymers by the same surfaces cannot be used to find the relationship between adhesion and adsorption. We shall explain this by a few examples.

Of the three polymers gelatin, polystyrene, and poly(methyl methacrylate), gelatin is known to have the highest adhesive strength during nonequilibrium disintegration. At the same time its adsorption from solutions on a glass surface is lowest. In fact, these data cannot be compared, since the adhesive strength here is due to other factors. The tensile strength in the system glass − gelatin − glass exceeds the strength of the glueing glass − polystyrene − glass, firstly because a weak boundary layer between the moist surface of the glass and the gelatin (hydrophilic polymer) is less probable than between the surface and a hydrophobic polymer and, secondly, because the cohesive strength of the gelatin is generally higher than that of polystyrene, and a mechanical disturbance of the polystyrene results in a cohesive breakaway. From the above example it follows that neither of the cases is directly related to adsorption.* Therefore, the data on adsorption can only be compared with the thermodynamically deduced work of adhesion. However, even in this case we encounter considerable difficulties that preclude any comparison.

In fact, both adsorption and adhesion (in equilibrium) depend on the type of reaction between the functional groups of the polymer and the surface, shape of the molecule, etc. However, the conditions for the formation of an adhesive bond differ greatly from those for polymer-adsorbent interaction in solution. During adsorption from solutions, the molecules of the polymer and the solution compete for sites on the surface, and this competition decreases the adsorption of the polymer and the strength of its attachment to the surface.

The discussions in the previous chapters show that adsorption is very dependent on the nature of the solvent, since this determines the shape of the chain, and thus the conditions of contact with the surface during adsorption. However, in the formation of an adhesive bond, these factors are almost always completely eliminated, even if the joint is applied by means of a solution. When polymers are applied in solvents that interact weakly with the surface, polymer adsorption is the primary act of the formation of a surface or adhesive film.

However, if the solvent actively interacts with the surface, the adhesive joint begins to be formed only from the moment when the greater part of the solvent is eliminated from the system, and a large number of bonds can be formed between the polymeric molecule and the surface when the functional groups of the polymer are no longer blocked by the solvent. When the solvent is removed during the formation of the film, the concentration of the solution near the surface gradually increases, and the ratio between the total number of interactions of the polymeric molecules and the solvent molecules sharply changes. At the same time, the structure of the polymer is also changed, and processes of the formation and relaxation of internal stresses occur, which affect the strength of the adhesive bond /242, 243/.

* The authors wish to thank Ya. O. Bikerman for his valuable suggestions.

Hence, the conditions under which polymers are adsorbed from solution, and the conditions of formation of an adhesive bond, differ greatly. This difference becomes even greater if the adhesive joint is not applied by means of a solution but in other ways. Therefore, we cannot experimentally establish a direct relationship between the adsorption of a polymer from a solution and its adhesion to the same surface, although it certainly exists. The character of the adsorption determines the structure of the film formed on the surface, which must affect the firmness of the adhesive bond. In particular, this refers also to the determination of the adhesive strength by nonequilibrium methods, since structural defects of the films are also related to their formation. We must distinguish here between the correlation between adsorption and adhesion (as the equilibrium characteristic determined by thermodynamic equations), and the correlation between adsorption and adhesive strength. In this sense adhesive strength is a nonequilibrium magnitude, determined by disturbance of the adhesive bond under nonequilibrium conditions.

EFFECT OF THE INTERFACE ON REACTIONS OF
SYNTHESIS AND THE STRUCTURE OF
THREE-DIMENSIONAL POLYMERS

The research of the school of Professor Kargin has shown that the processes of structure formations in polymers are relaxation processes depending on the molecular mobility of the structural chain elements. The processes of structure formation begin even during polymerization, and these processes are correlated /244/.

When polymeric materials with a given chemical and physical structure are produced, the preparation of reinforced plastics and filled polymers is of special importance. The processes of polymerization and structure formation take place simultaneously in the presense of a strongly developed surface of a fibrous or disperse filler. Kargin et al. /245/ studied the effect of small amounts of filler, that serve as nuclei in crystallizing polymers, on the crystallization processes.

However, up to now there have been few studies on the processes of structure formation in the presence of fillers, that is, on the effect of the interface on processes of polymerization and also structure formation. However, this problem is very important in the production of filled and reinforced polymers, when the processes of polymerization and structure formation occur near the polymer-solid interface. In some of our papers /23, 240, 259/ we studied the effect of the solid surface on the processes of structure formation during the molding of the polymeric material from the solution or the melt, and proved that the filler surface has an appreciable influence on these processes and the properties of polymers in the boundary layers.

Studies on the effect of the filler on the course of the reaction of formation of three-dimensional polymers and its properties are interesting.

We shall examine some results of research on this subject. They indicate the important role of the interface in processes of synthesis and structure formation. We studied the effect of the filler on the network structure of the three-dimensional copolymer of styrene and divinylbenzene. The monomers were copolymerized in the presence of finely disperse quartz powder /246/. We obtained copolymers with a content of divinyl-benzene of 3, 10, and 15 wt%, and a content of filler of 10, 30, 50, and 70 wt%. Thus, the total surface area of the solid on which polymerization was carried out was continuously increased.

To characterize the networks obtained, we used the swelling method and the Flory — Rehner theory. Table 27 contains data on the dependence of the swelling of the copolymers (with reference to the polymer without filler) in the presence of various amounts of filler for copolymers of different density of the chemical network. It can be seen that the intro-duction of a filler leads to considerable increase in swelling. Formal application of the Flory — Rehner theory to these data indicates that in the presence of filler the effective number of points in the three-dimensional network of the polymer decreases. At a first glance this is surprising, since we have already shown that the filler leads to the formation of additional points of cross-linkage /247/. However, it was found in /248, 249/ that in the presence of a filler an apparent increase in the swelling of vulcanized rubbers is observed, which involves impairment of the bond on the polymer — filler interface during swelling under the action of the solvent. This results in the formation of vacuoles filled by the solvent round the particle. In the case of complete impairment of the bond on the interface, the swelling of the filled polymer (with reference to the unfilled polymer) is determined by /250/:

$$Q = v_2^{-1} = (v_{r_0}^{-1} - \Phi)/(1 - \Phi), \qquad (7.14)$$

where v_2 is the volume fraction in the swollen sample; Φ is the volume fraction of the filler.

TABLE 27. Swelling of the copolymer of styrene and divinylbenzene in the presence of a filler in ethylbenzene

DVB, wt%	Content of filler, wt%								
	0	10		30		50		70	
	exp	exp	calc	exp	calc	exp	calc	exp	calc
3	2.48	3.07	2.55	3.20	2.80	3.50	3.20	5.68	4.5
10	1.84	2.10	1.88	2.30	2.02	2.73	2.25	2.65	2.8
15	1.68	1.82	1.67	2.28	1.78	2.68	1.94	3.19	2.3

Our calculation and determination of swelling in different solvents showed that in practically all cases the experimentally determined value Q is much larger than the value calculated from (7.14), assuming that the bonds are completely destroyed on the interface. It is quite clear that the effect of the sharp increase in the swelling of the systems studied involves changes in the density of the network formed by chemical bonds. This effect increases more at higher contents of the filler level in the polymer. It follows that during the formation of the three-dimensional network in polymerization in the presence of a filler, the process proceeds differently from that in the absence of a filler. Thus, we can assume that in the presence of a developed surface a more defective network is formed than during polymerization in the absence of a filler. The effect of the presence of a filler on the efficiency of cross-linking was noted for rubbers /250/, but here the already formed polymeric chains are cross-linked. In our case the polymer chain grows simultaneously with cross-linking.

The strongly developed surface of the filler in the initial stage of the reaction may lead to an increase in the rate of rupture of the reacting chains on the surface, so that the density of the network decreases and the network becomes more defective. Evidently, the surface of the filler plays the role of a special inhibitor in the formation of the three-dimensional network. The introduction of an inhibitor during polymerization at the beginning of network formation (during the increase in the viscosity of the system) led to an increase in the swelling of the three-dimensional polymer obtained in the absence of a filler for the same degree of conversion. Hence, because the inhibitor prevents the reaction of growth and cross-linking, it also leads to the appearance of a defective three-dimensional structure.

Apparently, in the later stages of the reaction another mechanism also exists which leads to an increase in the defectiveness of the network. Because of the adsorption of the growing polymeric chains on the filler surface, the mobility of the chains decreases sharply, which affects the rate of growth and of breaking. All these factors lead to the formation of a more defective structure of the three-dimensional network.

An experimental confirmation of the effect of the interface on the kinetics of formation of three-dimensional polymers can be shown by taking the kinetics of formation of three-dimensional polyurethans as an example /251/. The reaction kinetics of the formation of polyurethans was studied. These elastomers were prepared by cross-linking trimethylolpropane with macrodiisocyanates, obtained from poly(hydroxypropylene glycols), with molecular weights between 2000 and 1000, and also 4,4-diphenylmethane in the ratio of 1 : 2. The kinetics of polymer formation on a copper surface and in the bulk were studied by IR spectroscopy.

It was found that the curing rate on the surface is lower for the macro-diisocyanate on the base of a polyester with a lower molecular weight, while the rate is higher for the same macrodiisocyanate in the bulk. The behavior in the bulk agrees with the fact that the decrease in the molecular weight of the polyester leads to an increase in the concentration of the reactive isocyanate groups per unit volume. The anomalous occurrence of a reaction on the surface may be the result of the increase in the con-centration of the strongly polar NCO groups that appreciably promotes the reaction between the initial macrodiisocyanate and the polyurethan formed as the surface. This considerably reduces the mobility of the chains on the surface for the macrodiisocyanate based on the polyurethan with a lower molecular weight. However, in both cases the overall rate of the reaction on the surface is higher than in the bulk. This is explained by the fact that the adsorptional interaction of the macrodiisocyanates with the surface leads to a certain ordering of the molecules with reference to one another in the surface layer. According to Kargin and Kabanov /252/, this ordered structure leads to polymerization, and may accelerate the total rate of the process.

The interface thus has a double effect on the processes of synthesis and structure formation in three-dimensional polymers. It increases the probability of a breaking reaction in the first stages, and complicates the breaking in the higher reaction stages because of the adsorptional interaction of the growing chains with the surface. This in turn affects the reaction rate and the structure of the network. Thus, if we take into account the physical and chemical network points, we can assume that a significant characteristic of the network, such as the effective cross-link density, differs if the reaction is carried out in the presence or absence of an interface with the filler. This hypothesis is especially well illustrated by a study of a system in which the contribution of the physical network points to the effective density of the network is much greater than the contribution of the chemical points. Examples of such a system are three-dimensional polyurethans.

We studied the effective cross-link density of a polyurethan coating based on polyethers and polyesters and toluylene diisocyanate with differing ratios NCO/OH. The cross-linking agent was trimethylolpropane /254/. The effective cross-link density was determined for free films and films applied to aluminum surfaces from the data of the swelling by the Flory − Rehner method. The parameters of interaction of polymer − solvent systems were found. Table 28 shows the values found for M_c, the molecular weight of the chain section between the effective points of the network.

TABLE 28. M_c and v_c/V for polyurethan coatings

Polyester or polyether	NCO : OH	Free film		Film on surface	
		M_c	$v_c/V \cdot 10^4$	M_c	$v_c/V \cdot 10^4$
Diethylene glycol adipate,	2:1	620	21.0	210	61.0
molecular weight 800	1.75:1	970	13.4	130	100
	1.50:1	510	25.4	200	64.8
Poly (hydroxypropylene glycol),					
molecular weight 2000	2:1	520	21.3	460	23.7
750	2:1	290	43.1	200	60.6
550	2:1	120	109	180	75

The table shows that the effective network density for films present on the surface is usually higher than for free films. However, the variations in the density of the chemical network, given by ratio NCO/OH, and the physical (total) density of the network are not symbatic. This indicates a complex correlation between the chemical network structure, given by the number of functional groups taking part in the reaction, and the structure of the lattice given by the total number of junctions. In particular, the increase in the effective density of the network for films present on the surface indicates that the adsorption of growing chains on the surface during the reaction leads to the formation of additional junctions of the network. Their number, in turn, depends on the distance between the chemical junctions of the network. With increase in this distance, the flexibility of the chain section between these junctions increases, and the adaptability of the chain sections to the surface is higher. At the highest M_c in the free films, we observe the lowest M_c in the film on the surface (Table 28).

The differences in the chemical density of the network thus affect the properties of the films in the free state and on the surface. We should note that such effects could not be observed for the copolymers of styrene and divinylbenzene, where there are no functional groups able to interact with the surface. Table 28 also shows data on the effect of the molecular weight of the given polyester or polyether on the effective cross-link density for equal initial NCO/OH ratios. With increase in the molecular weight of the initial compound, the effective cross-link density decreases for both free films and films on the surface. This is due to the decrease in the total concentration of the functional groups that actively interact with the surface.

The presence of an interface thus introduces some special features into the formation of three-dimensional polymers. It affects the course of the chemical reaction and the effective cross-link density, and thus the polymer structure, which is very important for controlling the properties of surface films.

We believe that research in this field is necessary to establish the details of the mechanism of such effects by chemical and physical studies, and to relate them to the physical structure on the polymer surface. This is important for finding optimum ways of producing reinforced plastics and filled polymers, as for the above reasons the physicochemical properties and also the physicomechanical indexes of the boundary layers in these polymers change appreciably. A solution of the above problem is also important for finding general relationships between the chemical and structural transformations in polymers.

In conclusion we should note that the difference in the adsorbabilities of the reaction components may play an important part in the curing processes on the interface. If cured resin is a mixture of components, then the selective sorbability of any component by the mineral filler surface may be observed. This was shown by Trostyanskaya /255/ on the systems epoxy resins — polyamine, methylolphenols — phenol, and others. Thus, in the system epoxy resin — polyethylenepolyamine, the resin is preferentially adsorbed. The adsorbed phase does not participate in the curing reaction, while the resin phase, enriched by an excess of curing agent, becomes less rigid, since the polyethylenepolyamine that does not enter the reaction functions as a plasticizer. This reduces Young's modulus and increases the coefficient of thermal expansion of the resin phase. In the system methylololigomethylphenols — oligomethylenephenols the adsorbed film contains the preferentially adsorbed high-molecular weight compounds, and thus the structure of the cured resin phase is sharply changed.

From all these results it can be concluded that on the interface polymer — solid body, as on the interface polymer — gas, the molecular mobility of the polymeric chains appreciably decreases. This fact was proved by experiments on a large number of amorphous polymers by thermodynamic, structural, and mechanical methods, and is now considered to be well proved. However, the conclusion on the change in molecular mobility was the result of research on the properties of filled systems and coatings with properties that are generally determined by molecular mobility.

Characteristics of polymeric compounds such as transition temperature from one physical state into another, viscosity, relaxation characteristics, etc., correspond to the molecular mobility of chains, chain segments, and lateral groups. Therefore, all changes in the above characteristics that can be experimentally determined indicate a corresponding change in the molecular mobility. Finally, we could pass from indirect data to a direct determination of the molecular mobility of chains by the methods of NMR and dielectric relaxation, which confirm our earlier conclusions.

The change in molecular mobility leads to higher transition temperatures, especially glass points, changes in the crystallization conditions, and changes in the relaxation behavior of polymers in the boundary layers. In the last case this effect becomes apparent during formation of the polymeric material from the melt or solution, and during polymerization and the application of the prepared polymeric material.

The restriction in the molecular mobility in the boundary layers during polymer formation leads to inhibition of relaxation processes, and to the appearance of nonequilibrium stressed states, in comparison with the state of the polymer in the absence of a solid surface. As a result, loose molecular packing arises, and the filled polymer may have a lower average density (calculated on the polymer) than the unfilled one.

However, in the case of three-dimensional polymers, restriction of the molecular mobility during the synthesis of the network leads to removal of some of the molecular chains from the reactions of growth and formation of the network. Thus, the network formed will have more defects than that obtained in the absence of an interface. This general pattern was observed in numerous systems.

The decrease in the packing density, together with the restriction on the molecular mobility, changes the conditions for the occurrence of relaxation processes in the polymeric material, facilitates the processes involving the mobility of small chain elements as the result of the less dense packing, and inhibits the processes involved in the mobility of large structural elements. The relaxation spectrum expands. These consequences of the effect of the surface are the most important, although they are accompanied by changes in numerous other characteristics of the polymeric material.

We should bear in mind that here we are discussing the properties of the polymeric material, and not only the properties of the boundary layers. In the latter case we could not have determined characteristics such as are exhibited by a large number of molecules and are inherent in the polymeric substance. From this fact, and additional experimental data, it is possible to assume that the effect of the surface propagates over large distances. The far-acting effect of the surface forces is not, of course, a direct result of the field of force of the surface, but it is the result of the total change in the intermolecular actions in the system, and a change in the interactions between the chains that are in direct contact with the surface-adjacent chains.

A specific feature of polymers, namely, strong interaction between the polymeric molecules, thus leads to the propagation of the surface effect into the bulk. In fact, we can consider that molecular aggregates or other supermolecular structures participate in the interaction with the surface. The conclusion that molecular aggregates and not isolated molecules participate in the interaction events explains the remote effect of the surface, and also follows from our adsorption studies. A restriction on the mobility of even a single molecule of the aggregate leads to changes in the behavior of all the molecules of a given aggregate.

Finally, what is the reason for the restriction on the molecular mobility near the interface? The simplest answer to this question would be the hypothesis of purely adsorptional interaction between molecule and surface. This can be confirmed by numerous data on the effect of the surface on the adhesive, mechanical, and other properties of the boundary layers of polymers. In fact, adsorptional interaction is one of the two significant factors that determine the changes in the molecular mobility of the chains near the interface. It determines, in particular, the adhesion and the strength properties of filled and reinforced systems and glues.

However, there is also another factor that changes the mobility, namely, a purely entropic one that is not related to energetic interaction between polymer and surface. Calculations show that the molecule cannot have the same number of conformations at the interface as in the bulk, since the surface imposes restrictions on the geometry of the molecules. Therefore, the number of states of the molecules in the boundary layer decreases, together with its entropy, which is kinetically equivalent to a lower molecular mobility. The role of the entropy factors is confirmed by our calculations on the enthalpy and entropy of activation of the relaxation processes in the boundary layers. Theoretically, the same change in the molecular mobility in the boundary layers is possible if the energy of interaction between polymer and surface is the same. In this case we may observe comparable changes in the parameters related to molecular mobility for greatly differing mechanical properties determined by the strength of adhesion of the polymeric molecule to the surface.

The data of this chapter thus indicate the important role of adsorptional interaction on the polymer − solid interface in processes for preparing and applying heterogeneous polymeric materials. A study of the adsorption of polymers from solutions can clarify the patterns of such an interaction, and thus lead to a better understanding of the processes occurring in real systems that are important in practice.

CONCLUSION

From an analysis of the material of this monograph it is possible to form some general conclusions on the adsorption of polymers on solid surfaces. These conclusions are based on the modern theory of dilute polymer solutions and the conformational statistics of polymeric chains. If we take into account the behavior of macromolecules in dilute solutions, based on statistical thermodynamics, we can establish the main pattern of the adsorption of polymers, and its dependence on the nature of the polymer, the surface, the molecular distribution of the polymer, the nature of the solvent, and the temperature.

From the results of adsorption measurements, some fundamental conclusions can be derived on the character of the attachment of polymeric molecules to the adsorbent surface, and their distribution over it, and hence also on the structure of the adsorbed film. The theoretical description of adsorption within the framework of conformational statistics of the polymers has to a certain extent made it possible to forecast the behavior of polymeric molecules when they are adsorbed by a solid surface. Overall concepts on the mechanism of the adsorption processes are very important for solving general problems of these surface phenomena, where adsorption predominates.

However, it is evident that the theory of adsorption and adsorptional interaction of polymeric molecules with the surfaces of solids is still insufficiently developed for the reliable prediction of adsorption and an explanation of many experimental facts. We shall deal with some actual problems of the theory of adsorption. Their formulation must be based on established experimental facts, which are available in a sufficient amount.

The main task of the theory of adsorption is to set up an equation connecting adsorption with the polymer concentration in the solution or, in other words, the adsorption isotherm. All the existing theories on adsorption considered in Chapter 5 are applicable over a restricted range, or are difficult to verify experimentally. Thus, all these theories are less important for a quantitative description of adsorption, but can describe qualitative adsorption patterns, which are, however, based on a strictly mathematical consideration of the problem.

In fact, the basis of the statistical formulation is the concept of lattice types, that describe both the solution of the polymer and the surface of the adsorbent. However, the parameters are selected from general consider- ations and not from information on actual systems. The results may differ, depending on the premises, and thus a comparison with experimental data is very difficult, since many parameters that enter the theoretical equations cannot be experimentally determined. Hence, the concept on the structure of the adsorbed film is to a certain extent dependent on the model and the mathematical methods of calculation.

Considerable theoretical difficulties arise when the parameters characterizing the flexibility of the polymeric chain are introduced into the theory. In most cases parameters of chain flexibility are introduced that cannot be experimentally determined or compared with the parameters adopted in the theory of polymeric solutions.

In the thermodynamic approach to the derivation of the equation of the adsorption isotherm, there are appreciable difficulties in the determination of the partition functions of the adsorbed chain. This is because we do not yet have sufficient experimental data on the conformations of polymeric chains on the interface. In fact, conclusions on the structure of the adsorbed layer and the conformations of the adsorbed chains, as shown by the data given in Chapter 4, are exclusively based on indirect data or on experimental results obtained by using theoretical equations. In dilute solutions the conformations of macromolecules and their variation under the effect of different factors can be experimentally determined, but this cannot yet be done for molecules adsorbed by the surface.

However, the problem of the change in chain conformation near the interface is the main problem involved in an examination of the structure of adsorbed films, and it cannot be solved by theoretical calculations alone. When we consider the problem of conformation of chains in the boundary layer, we must bear in mind that its changes may be caused by both interaction with the surface, accompanied by changes in enthalpy, and the entropy factor, as the result of which the molecules near the interface cannot take up the same number of conformations as they can in the bulk.

In any consideration of the structure of the adsorbed film with complete surface coverage, we must take into account the interaction of the adsorbed molecules with one another, as this will also affect the conformation of the molecular chains in the adsorbed film. This has not hitherto been carried out in practice, although there are indications on the necessity for allowing for this factor.

The existing concept on the structure of the adsorbed film leads to another additional difficulty. This consists in the fact that the adsorbed film is considered to be a solution of the polymer, at a concentration much higher than that of the polymer in the bulk phase. Then, both the thermodynamic and statistical descriptions of the behavior of macromolecules must differ from those adopted for a dilute solution. In this case the possible aggregation of macromolecules in the adsorbed film must be considered. The introduction of concepts on the aggregation of macromolecules in the adsorbed film, complicated by the effect of the surface, is a condition for further successful development of the theory of adsorption.

The problem of multilayer adsorption is still unclear. In essence, a theory of multilayer adsorption does not exist, although many experimental data indicate that adsorption cannot be monomolecular. Evidently, multilayer adsorption cannot be described within the framework of a simple model which considers that monomolecules connect successive layers of macromolecules with the first adsorbed layer.

All the difficulties and problems that arise when we consider the theory of adsorption that we have indicated refer to the theory of adsorption from dilute solutions. Transition to more concentrated systems leads to additional difficulties. In this case, changes in the concentration of the solution lead to changes in the conformation of the macromolecules and the conditions of their interaction. The appearance of molecular aggregates which, as shown in many papers, begins even in dilute solutions, leads to the fact that at each concentration magnitude we deal with adsorbed particles that differ from one another in both shape and size. Hence, the conditions of contact of the molecules and the aggregates with the adsorbent surface, and hence the structure of the adsorbed film, change. We assume that further development of the theory of adsorption is impossible without allowance for these changes in the structure of solutions that occur when their concentration increases.

To solve the problem of surface phenomena in polymers, and especially the problems involved in adhesion, we must also study the conditions of adsorptional interaction of polymeric molecules with the surface in very concentrated systems or in the absence of a solvent. Research in this direction has not yet been initiated.

It follows from the above that in the theory of adsorption there are still many unsolved problems, which are important for a correct understanding of the process mechanism. It should, however, be borne in mind that not all the unsolved problems are theoretical. The theory develops slowly because we have insufficient experimental data on some problems. These include an experimental study of the effect of the flexibility of the polymeric chain under conditions when the energy of interaction with the surface is maintained constant. There are no direct determinations of the energy of adsorption of polymers on solid surfaces, and the effect of polydispersity on adsorption has been insufficiently studied. There are no data on the adsorption of block and graft polymers, which could give us information on the conditions of adsorption interaction, the problems involved in the adsorption of crystallizable polymers and oligomers have not been treated, and so on. There are no direct experimental data on the structure of the adsorbed film. This list of unsolved problems could be continued. The solution of the problems involved in adsorption is of theoretical and practical importance. Strictly speaking, the problem itself arises from practical necessity.

We shall finish this book with the words of one of the founders of modern physical chemistry, W. Ostwald: "There is no science for the sake of science... There is only science for the fulfillment of the purposes of mankind.... That which today is a purely abstract scientific problem, may tomorrow become the basis of an important technical problem."

BIBLIOGRAPHY

1. Lipatov, Yu. S. — Vestn. AN UkrSSR 9 (1970), 38.
2. Lipatov, Yu. S. — In: VI Yubileinaya Vsesoyuznaya konferentsiya po kolloidnoi khimii, p.54. Synopses of Reports. Izd. Voronezhsk. Univ., 1968. (Russian)
3. Deryagin, B. V. and N. A. Krotova. Adhesion. — Moscow, Izd. AN SSSR, 1949. (Russian)
4. Krotova, N. A. Gluing and Adhesion. — Moscow, Izd. AN SSSR, 1960. (Russian)
5. Voyutskii, S. S. Adhesion and Autohesion of Polymers. — Moscow, Rostekhizdat, 1963. (Russian)
6. Houwink, R. (Ed.). Adhesion and Adhesives. — London, Elsevier Publ. Co., 1957.
7. Adhesion. — Moscow, IL, 1954.
8. Moskvitin, P. I. Gluing of Polymers. — Lesnaya Promyshlennost', Moscow, 1968. (Russian)
9. McLaren, A. D. — J. Polymer Sci. 3 (1958), 652.
10. Berlin, A. A. and V. E. Basin. Principles of the Adhesion of Polymers. — Moscow, "Khimiya," 1969. (Russian)
11. Sharpe, L. H. and H. Schonhorn. — Chem. Eng. News 41 (1963), 41.
12. Bikerman, J. J. — Ind. Eng. Chem. 59 (1967), 41.
13. Bikerman, Ya. O. — Vysokomol. Soed. A10 (1968), 974.
14. Voyutskii, S. S. and B. V. Deryagin. — Kolloid. Zh. 27 (1965), 624.
15. Zisman, W. — Ind. Eng. Chem. 55 (1963), 19.
16. Shafrin, E. and W. Zisman. — J. Phys. Chem. 64 (1960), 519.
17. Ber, E. (Ed.). Structural Properties of Plastics. — Moscow, "Khimiya" (1967), 274. (Russian)
18. Lipatov, Yu. S. Physics and Chemistry of Filled Polymers. — Kiev, "Naukova Dumka," 1967. (Russian)
19. Lipatov, Yu. S. and F. G. Fabulyak. — International Symposium on Macromolecular Chemistry. Preprint A12—6. Toronto, 1968.
20. Lipatov, Yu. S. and F. G. Fabulyak. — Vysokomol. Soed. A10 (1968), 1952.
21. Lipatov, Yu. S. and F. G. Fabulyak. — Vysokomol. Soed. B12 (1970), 871.
22. Lipatov, Yu. S. and L. M. Sergeeva. — Vysokomol. Soed. 8 (1966), 1895.
23. Lipatov, Yu. S. — Vysokomol. Soed. A10 (1968), 2737.
24. Wolfram, F. — Koll. Z. u Z. Polymere 133 (1962), 439.
25. Kroser, S. — J. Polymer Sci. 182 (1962), 75.
26. Fainerman, A. E., Yu. S. Lipatov, and V. M. Kulik. — Kolloid. Zh. 31 (1969), 142.
27. Fainerman, A. E. and Yu. S. Lipatov. — In: Surface Phenomena in Polymers, p. 19. Kiev, "Naukova Dumka," 1970. (Russian)
28. Priel, Z. and A. Silberberg. — Amer. Chem. Soc. Polymer Preprints 11 (1970), 1405.
29. Harkins, W. D. The Physical Chemistry of Surface Films. — New York, Reinhold Publ. Corp., 1952.
30. Davis, J. and E. Rideal. Interfacial Phenomena. — New York, Academic Press, 1961.
31. Fainerman, A. E., Yu. S. Lipatov, and V. K. Maistruk. — DAN SSSR 188 (1969), 152.
32. Fainerman, A. E., Yu. S. Lipatov, and V. K. Maistruk. — DAN SSSR 178 (1968), 1129.
33. Yamashita. — Bull. Chem. Soc. Japan 38 (1965), 430.
34. Fainerman, A. E., Yu. S. Lipatov, and V. K. Maistruk. — Kolloid. Zh. 32 (1970), 382.
35. Tsvetov, V. P., V. E. Eskin, and S. Ya. Frenkel'. The Structure of Macromolecules in Solution. — Leningrad, "Nauka," 1964. (Russian)
36. Marcelin, A. Surface Solutions. — Kolloidbeihefte 38 (1933), 177—336.
37. Lipatov, Yu. S., N. G. Peryshkina, and L. M. Serveeva. — DAN SSSR 6 (1962), 42.
38. Lipatov, Yu. S. and L. M. Sergeeva. — Kolloid. Zh. 27 (1965), 217.
39. Patat, F., E. Killmann, and C. Schliebener. — Fortschr. Hochpolym. Forsch. 3 (1964), 332.
40. Kraus, G. — Rubb. Chem. Technol. 38 (1965), 1070.

41. Kiselev, B. A. Glass-Fiber Reinforced Plastics. — Moscow, Goskhimizdat, 1961. (Russian)
42. Andreevskaya, G. D. High-Strength Oriented Glass-Fiber Reinforced Plastics. — Moscow, "Nauka," 1966. (Russian)
43. Schwartz, R. (Ed.). Fundamental Aspects of Fiber-Reinforced Plastic Composites. — New York—London, Interscience Publ., 1968.
44. Tikhomirov, V. B. Physicochemical Principles for Producing Nonwoven Materials. — Legkaya Industriya, Moscow, 1969. (Russian)
45. Stromberg, R. R., A. R. Quasius, C. Toner, and M. Parker. — J. Res. Nat. Bur. Stand. 62 (1959), 71.
46. Kiselev, A. V., V. I. Lygin, and I. N. Solomonova. — Kolloid. Zh. 30 (1968), 386.
47. Mizukara, K., K. Hara, and T. Imoto. — Koll. Z. u. Z. Polymere 229 (1969), 17.
48. Stromberg, R. R., W. H. Grant, and E. Passaglia. — R. Res. Nat. Bur. Stand. (Phys. a. Chem.), 68A (1964), 391.
49. Stromberg, R. R. — Kunststoffe 12 (1965), 12.
50. Sato, T., T. Tanaka, and T. Yoshida. — J. Polymer Sci. B5 (1967), 947.
51. Growl, W. — J. Oil and Colour Chem. Assoc. 46 (1963), 204.
52. Tolstaya, C. H. Author's Summary of Doctoral Thesis. — IFKh AN SSSR, Moscow, 1969. (Russian)
53. Felter, R. E., E. S. Moyer, and Z. N. Ray. — J. Polymer Sci. B7 (1969), 533.
54. Perkel, R. and R. Ullman. — J. Polymer Sci. 54 (1961), 127.
55. Klenin, V. I. — In: The Mechanism of Processes for the Formation of Films from Polymeric Solutions and Dispersions, p. 32. Moscow, "Nauka," 1966. (Russian)
56. Heller, L. W., H. L. Bhatnager, and M. Nakagaki. — J. Chem. Phys. 36 (1962), 1163.
57. Alekseev, A. M., V. D. Fikhman, and V. I. Klenin. — Vysokomol. Soed. A12 (1970), 2532.
58. Lipatov, Yu. S., T. T. Todosiichuk, and L. M. Sergeeva. — DAN UkrSSR B4 (1971), 333.
59. Lipatov, Yu. S., T. T. Todosiichuk, and L. M. Sergeeva. — Vysokomol. Soed. B13 (1972), 2542.
60. Grakin, F. M., E. Passaglia, R. R. Stromberg, and H. Steinberg. — J. Res. Nat. Bur. Stand. 67A (1963), 363.
61. Killman, E. and H. Welgand. — Macromol. Chem. 132 (1970), 239.
62. Stromberg, R., D. Tutas, and E. Passaglia. — J. Phys. Chem. 69 (1965), 3955.
63. Peyser, P. and R. Stromberg. — J. Phys. Chem. 71 (1967), 2066.
64. Fontana, B. and J. Thomas. — J. Phys. Chem. 65 (1961), 480.
65. Peyser, P., D. Tutas, and S. Stromberg. — J. Polymer Sci. A5 (1967), 653.
66. Öhrn, O. — Arkiv kemi 12 (1958), 397; Macromol. Chem. 18—19 (1956), 383.
67. Hugue, M., F. Fischman, and D. Goring. — J. Phys. Chem. 63 (1959), 766.
68. Doroszkowski, A. and R. Lambourne. — J. Colloid Interface Sci. 26 (1968), 214.
69. Malinskii, Yu. M. — Usp. Khim. 39, No. 8 (1970), 1511.
70. Sagalaev, G. V. — In: Fillers of Polymeric Materials, p. 18. Moscow, Izd. Doma Nauchno-Tekhn. Propagandy im. Dzerzhinskogo, 1969. (Russian)
71. McCabe, K. — In: Kraus, G. Reinforcement of Elastomers. New York, Wiley, 1965.
72. Bogacheva, E. K. and Yu. A. El'tekov. — In: Surface Phenomena in Polymers, p. 52. Kiev, "Naukova Dumka," 1970. (Russian)
73. Stromberg, R. R. and G. M. Kline. — Modern Plastics 4 (1961), 20.
74. Hobden, J. F. and H. H. Jellinek. — J. Polymer Sci. 11, No. 4 (1953), 365.
75. Bogacheva, E. K., A. V. Kiselev, Yu. S. Nikitin, and Yu. A. El'tekov. — Vysokomol. Soed. 10A, No. 3. (1968), 574.
76. Kraus, G. and I. Dugone. — Ind. Eng. Chem. 47 (1955), 1809.
77. Koral, L, R. Ullman, and F. Eirich. — J. Phys. Chem. 62 (1958), 541.
78. Burns, H. and D. Carpenter. — Amer. Chem. Soc. Polymer Preprints 5 (1964), 517.
79. El'tekov, Yu. A. — In: Surface Phenomena in Polymers, p. 45. Kiev, "Naukova Dumka," 1970. (Russian)
80. Thies, C. — J. Phys. Chem. 70, No. 12 (1966), 3783.
81. Peterson, C. and T. Kwei. — J. Phys. Chem. 65 (1961), 1331.
82. Shyluk, W. P. — J. Polymer Sci. 6 (1968), 2009.
83. Binford, G. S. and E. M. Gessler. — J. Phys. Chem. 63 (1959), 1376.
84. Howard, G. and P. McConnel. — J. Phys. Chem. 71 (1967), 2974, 2981, 2991.
85. Claesson, J. and S. Claesson. — Arkiv kemi, min. geol. A19 (1944—45), 1.

86. Kraus, G. and J. T. Gruver. — Rubb. Chem. Techn. **41** (1968), 1256.
87. Arendt, O. — Koll. Beinhefte **7** (1915), 212.
88. Patat, F. and C. Schliebener. — Macromol. Chem. **44—46** (1961), 643—668.
89. Kangle, P. and E. Pacsu. — J. Polymer Sci. **54** (1961), 301.
90. Heller, W. and W. Tanaka. — Bull. Amer. Phys. Soc. **26** (1961), 17.
91. Kolthoff, I. M. and R. G. Gutmacher. — J. Phys. Chem. **56** (1952), 740.
92. Nakato Kanzi, Nobuo Shiranshi, and Kuniharu Yoko. — J. Soc. Mater. Sci. Japan **16** (1967), 839.
93. Ermilov, P. I. — Lakokras. Mater. i Ikh Prim. **1** (1969), 18.
94. Rapchinskaya, S. E. and G. A. Blokh. — In: Surface Phenomena in Polymers, p. 70. Kiev, "Naukova Dumka," 1970. (Russian)
95. Gottlieb, M. — J. Phys. Chem. **64** (1960), 427.
96. Si Jung Ye, and H. L. Frisch. — J. Polymer Sci. **27** (1958), 149.
97. Yurzhenko, A. I. and I. I. Maleev. — J. Polymer Sci. **31** (1958), 301.
98. Frisch, H., M. J. Hellman, and J. L. Lundberg. — J. Polymer Sci. **38** (1959), 444.
99. Killman, E. and E. Schneider. — J. Macromol. Chem. **57** (1962), 212.
100. Jankovics, L. — J. Polymer Sci. **A3** (1965), 3519.
101. La Mer, V. K. and R. H. Smeller. — J. Coll. Sci. **13** (1958), 589.
102. Kipling, R. Adsorption from Solution of Nonelectrolytes, p. 70. — London, Acad. Press, 1965.
103. Flory, P. Principles of Polymer Chemistry, New York, 1953.
104. Schulz, G. V. — Angew. Chem. **64** (1952), 553.
105. Schulz, G. V. and U. Kantow. — J. Polymer Sci. **10** (1953), 79.
106. Kolthoff, J. M., R. G. Gutmacher, and A. Kahn. — J. Phys. Chem. **55** (1951), 1240.
107. Gilliland, E. R. and E. B. Gutoff. — J. Appl. Polymer Sci. **3** (1960), 26.
108. Mark, H. and G. Saito. — Monatsh. **68** (1936), 237.
109. Brooks, M. C. and R. M. Badyer. — J. Amer. Chem. Soc. **72** (1950), 4384.
110. Jenkel, F. and B. Rumbach. — Z. Elektrochem. **55** (1951), 612.
111. Hara, K. and T. Imoto. — Koll. Z. u. Z. Polymere **237** (1970), 297.
112. Ellerstein, S. and R. Ullman. — J. Polymer Sci. **55** (1961), 161, 123.
113. Luce, I. E. and A. A. Robertson. — J. Polymer Sci. **51** (1961), 155, 317.
114. Gregg, S. I. and J. Jacobs. — Trans. Farad. Soc. **44** (1948), 574.
115. Soltys, M. N., I. I. Maleev, T. M. Polonskii, and I. I. Mikityuk. — In: Surface Phenomena in Polymers, p. 65, Kiev, "Naukova Dumka," 1970. (Russian)
116. Skrylev, L. D. and R. E. Savina. — Kolloid. Zh. **27**, No. 4 (1965), 605.
117. Bothman, R. and C. Thies. — J. Coll. Interface Sci. **31** (1969), 1.
118. Tul'bovich, B. I. and E. I. Priimak. — Zh. Fiz. Khim. **43**, No. 4 (1969), 960.
119. Zakordonskii, T. M., T. M. Polonskii, et al. — In: Surface Phenomena in Polymers, p. 58, Kiev, "Naukova Dumka." 1970. (Russian)
120. Stromberg, R. and G. Kline. — Poliplasti **9** (1961), 15.
121. Polonskii, T. M. and V. P. Zakondorskii. — Teor. Eksp. Khim. **6** (1970), 367.
122. Tolstaya, S. I., V. N. Borodina, and A. B. Taubman. — Kolloid. Zh. **27** (1965), 446.
123. Sonntag, F. and J. Jenkel. — Koll. Z. **135** (1954), 81.
124. Parfitt, R. L. and D. Greenland. — Clay Miner. **8** (1970), 305.
125. Claesson, S. — Disc. Farad. Soc. **7** (1949), 34.
126. Ingelman, B. and M. Halling. — Arkiv kemi **1** (1949—50), 61.
127. Golub, M. A. — J. Polymer Sci. **11** (1953), 583.
128. Sen'kin, N. P. Author's Summary of Candidate Thesis. — L'vovskii Universitet, 1970. (Russian)
129. Miller, B. and E. Pacsu. — J. Polymer Sci. **41** (1959), 87.
130. Mizuhara, K., K. Hara, and T. Imoto. — Koll. Z. u. Z. Polymere **238** (1970), 442.
131. Roe, R. — Proc. Natl. Acad. Sci. (U.S.) **53** (1965), 50.
132. Maleev, I. I., T. M. Polonskii, and M. N. Soltys. — Vysokomol. Soed. **10A**, No. 9 (1968), 2122.
133. Kiselev, A. V., Yu. A. El'tekov, and E. K. Bogacheva. — Kolloid. Zh. **26**, No. 4 (1964), 458.
134. Kiselev, A. V., V. N. Novikov, and Yu. A. El'tekov. — DAN SSSR **149**, No. 1 (1963), 131.

135. Kiselev, A. V., Yu. A. El'tekov, and V. N. Novikova. — In: The Mechanism of Processes for the Formation of Films from Polymer Solutions and Dispersions, p. 85. Moscow, "Nauka," 1966. (Russian)

136. Trapeznikov, A. A. et al. — DAN SSSR 160, No. 1 (1965), 174.

137. Kiselev, A. V., N. V. Kovaleva, and A. L. Korolev. — Kolloid. Zh. 23 (1961), 583.

138. Malysheva, L. N., A. B. Rabinovich, M. C. Kravchenko, and N. N. Dmitrieva. — In: Petroleum Processing, 6, p. 9—11. TsNIITEneftegaz, Moscow, 1969. (Russian)

139. Tolstaya, S. N. et al. — In: Detailed Summary of the Proceedings of the Second All-Union Conference on Theoretical Problems of Adsorption, p. 53. Moscow, "Nauka," 1969. (Russian)

140. Tolstaya, S. N., S. S. Mikhailova, A. B. Taubman, and A. V. Uvarov. — DAN SSSR 178, No. 1 (1968), 148.

141. Erman, V. Yu., S. N. Tolstaya, and A. B. Taubman. — Kolloid. Zh. 31, No. 4. (1969), 617.

142. Tolstaya, S. N. et al. — Lakokras. Mat. i Ikh Prim., No. 6 (1967), 8—10.

143. Tolstaya, S. N. and S. A. Shabanova. — Lakokras. Mat. i Ikh Prim., No. 1 (1969), 6—8.

144. Tolstaya, S. N. et al. — Lakokras. Mat. i Ikh Prim. No. 6. (1966), 45—48.

145. Shabanova, S. A. Author's Summary of Candidate Thesis. — IFKh AN SSSR, Moscow, 1965. (Russian)

146. Di Marzio, E. A. — J. Chem. Phys. 42 (1965), 2101.

147. Thies, C. — Macromol. 1 (1968), 335.

148. Bothman, R. and C. Thies. — Preprints International Symposium on Macromolecular Chemistry, Preprint A7—6, Toronto, 1968.

149. Silberberg, A. — Amer. Chem. Soc. Polymer Preprints 11 (1970), 1256.

150. Schick, M. J. and E. N. Harvey. — J. Polymer Sci. B7 (1969), 495.

151. Uvarov, A. R., R. Yu. Erman, and N. A. Aleksandrova. — Kolloid. Zh. 32, No. 5 (1970), 791.

152. Thies, C. — Amer. Chem. Soc. Polymer Preprints 6 (1965), 320.

153. Thies, C. — Amer. Chem. Soc. Polymer Preprints 7 (1966), 880.

154. Di Marzio, E. A. and F. L. McCrackin. — J. Chem. Phys. 43 (1965), 539.

155. Fendler, H., H. Rohlender, and U. Stuart. — Macromol. Chem. 18/19 (1956), 383.

156. El'tekov, Yu. A. — In: Detailed Summary of the Proceedings of the Second All-Union Conference on Theoretical Problems of Adsorption, Moscow, "Nauka," 1969. (Russian)

157. Rowland, F., R. Bulas, E. Rothstein, and F. Eirich. — Ind. Eng. Chem. 57, No. 9 (1965), 49.

158. Rowland, F. and F. Eirich. — J. Polymer Sci. A1 (1966), 2401.

159. Bulas, R. Thesis. — Polytech. Inst. Brooklyn, 1963.

160. Malinskii, Yu. M. — Vysokomol. Soed. 8, No. 11 (1966), 1886.

161. Lipatov, Yu. S., L. M. Sergeeva, and V. P. Maksimova. — Vysokomol. Soed. 2 (1960), 1569.

162. Di Marzio, E. D. — [Ref. incorrect].

163. Stromberg, R. and L. Smith. — J. Phys. Chem. 71, No. 8 (1967), 2470.

164. Silberberg, A. — Amer. Chem. Soc. Polymer Preprints 11 (1970), 1289.

165. Rubin, R. — J. Res. Nat. Bur. Stand. B69 (1965), 301.

166. Silberberg, A. — J. Phys. Chem. 66 (1962), 1872.

167. Steele, W. — In: Interphase boundary, gas—solid. — Moscow, "Mir," 1970. (Russian)

168. Hobden, I. and H. Ellinek. — J. Polymer Sci. 11 (1953), 365.

169. Bindorf, I. and E. Gessler. — J. Phys. Chem. 63 (1959), 1376.

170. Gyani, B. — Ind. Eng. Chem. 21 (1944), 79.

171. Treiber, E., C. Porod, W. Gierlinger, and G. Schulz. — Macromol. Chem. 9 (1953), 241.

172. Simha, R., H. Frisch, and F. Eirich. — J. Phys. Chem. 57 (1953), 584.

173. Simha, R., H. Frisch, and F. Eirich. — J. Chem. Phys. 25 (1953), 365.

174. Frisch, H. and R. Simha. — J. Phys. Chem. 58 (1954), 507.

175. Frisch, H. — J. Phys. Chem. 59 (1955), 633.

176. Simha, R., H. Frisch, and F. Eirich. — J. Polymer Sci. 29 (1958), 3.

177. Frisch, H. and R. Simha. — J. Chem. Phys. 27 (1957), 702.

178. Fontana, B. — J. Phys. Chem. 70 (1966), 1801.

179. Silberberg, A. — J. Phys. Chem. 66 (1962), 1884.

180. Silberberg, A. — J. Chem. Phys. 46 (1967), 1105.

181. Silberberg, A. — J. Chem. Phys. **48** (1968), 2835.
182. Rubin, R. — J. Chem. Phys. **43** (1965), 2392.
183. Roe, R. — J. Chem. Phys. **43** (1965), 1591.
184. Hoeve, C. — J. Chem. Phys. **43** (1965), 3007.
185. Hoeve, C. — J. Chem. Phys. **44** (1966), 1505.
186. Hoeve, C., E. Di Marzio, and P. Peyser. — J. Chem. Phys. **42** (1965), 2558.
187. Hoeve, C. — Amer. Chem. Soc. Polymer Preprints **11** (1970), 1232.
188. Hoeve, C. — Preprints International Symposium on Macromolecular Chemistry, Preprint A7—1, Toronto, 1968.
189. Rubin, R. — J. Res. Nat. Bur. Stand. **B70** (1966), 237.
190. Pouchly, J. — Preprints International Symposium on Macromolecular Chemistry, Preprint A7—2, Toronto, 1968.
191. Hoffman, R. and W. Torsman. — Ibid., Preprint A7—4.
192. Tarsman, W. and R. Huges. — J. Chem. Phys. **38** (1963), 2310.
193. Di Marzio, E. and R. Rubin. — Amer. Chem. Soc. Polymer Preprints **11** (1970), 1239.
194. Simcha, R. and J. Zakin. — J. Chem. Phys. **33** (1960), 1791.
195. Lipatov, Yu. S. Author's Summary of Doctoral Thesis. — FKhI im. Karpova, Moscow, 1963. (Russian)
196. Lipatov, Yu. S. and N. F. Proshlyakova. — Usp. Khim. **30** (1961), 513.
197. Slonimskii, G. L. et al. — Vysokomol. Soed. **BX**, No. 9 (1968), 640.
198. Patat, F. and L. Estupian. — Macromol. Chem. **49** (1961), 182.
199. Lipatov, Yu. S., V. P. Maksimova, and L. M. Sergeeva. — Vysokomol. Soed. **2** (1960), 596.
200. Lipatov, Yu. S., N. G. Peryshkina, and L. M. Sergeeva. — Vysokomol. Soed. **4** (1962), 596.
201. Lipatov, Yu. S. and L. M. Sergeeva. — In: Ion Exchange and Sorption from Solutions, p. 63. Minsk, Izd. AN BSSR, 1963. (Russian)
202. Lipatov, Yu. S. — In: Adhesion of Polymers, p. 63. Moscow, Izd. AN SSSR, 1963. (Russian)
203. Tager, A. A. — Usp. Khim. **27** (1958), 481.
204. Schutte, H. — Plaste u. Kautschuk **11** (1964), 248.
205. Crowl, V. T. and M. A. Malai. — Disc. Farad. Soc. **42** (1966), 301.
206. Porowska, E. and Z. Hippe. — Rocz. Chem. **44** (1970), 635.
207. Aripov, E. A. et al. — Uzb. Khim. Zh. **6** (1968), 26.
208. Makhkamova, K. M. and K. S. Akhmedov. — DAN UzbSSR **5** (1966), 31—34.
209. Kargin, V. A., Z. A. Berestneva, and M. B. Konstantinopol'skaya. — Vysokomol. Soed. **1** (1959), 1074.
210. Lipatov, Yu. S., P. I. Zubov, and E. A. Andryushchenko. — Kolloid. Zh. **21** (1959), 598.
211. Belen'kii, V. G. and E. S. Gankina. Thin-Layer Chromatography of Polymers. — Izd. IVS AN SSSR. Leningrad, 1970. (Russian)
212. Zubov, P. N., Yu. S. Lipatov, and E. A. Kanevskaya. — DAN SSSR **141** (1961), 387.
213. Lipatov, Yu. S. and L. M. Sergeeva. — Kolloid. Zh. **27** (1965), 217.
214. Simha, R. — J. Polymer Sci. **29** (1958), 3.
215. Trapeznikov, A. A. — DAN SSSR **160** (1965), 174.
216. Rusanov, A. I. Phase Equilibria and Surface Phenomena. — Leningrad, "Nauka," 1968. (Russian)
217. Lipatov, Yu. S. — Vysokomol. Soed. **7** (1965), 1430.
218. Lipatov, Yu. S. and F. G. Fabulyak. — Vysokomol. Soed. **AX** (1968), 1592.
219. Lipatov, Yu. S. and F. G. Fabulyak. — Vysokomol. Soed. **AX 1** (1969), 724.
220. Lipatov, Yu. S. and T. E. Geller. — Vysokomol. Soed. **8** (1966), 582.
221. Lipatov, Yu. S. and T. E. Geller. — Vysokomol. Soed. **9A** (1967), 241.
222. Lipatov, Yu. S. — Vysokomol. Soed. **BX** (1968), 527.
223. Braun, W. F. Jr. Dielectrics. — In: Flügge, S. (Ed.). Handbuch der Physik, Vol. 17, p. 1. Berlin, Springer Verlag, 1956.
224. Cole, H. and K. Cole. — J. Chem. Phys. **9** (1941), 341.
225. Frisch, H. and S. A. Madfai. — J. Amer. Chem. Soc. **80** (1958), 3561.
226. Higushi, W. — J. Chem. Phys. **65** (1961), 487.

227. Roe,R., D. Davis, and K. Kwei. Papers Presented at Chicago Meeting of Amer. Chem. Soc.
      Div. of Organic Coatings and Plastics Chem. Sept.,1970.
228. Waldrop,M. and G. Kraus. — Rubb. Chem. Techn. 42 (1969), 1155.
229. Lipatov,Yu. S. and T. E. Lipatova. — Vysokomol. Soed. 5 (1963), 290.
230. Lipatov,Yu. S. and L. M. Sergeeva. — DAN BSSR 8 (1964), 594.
231. Lipatov,Yu. S. and L. M. Sergeeva. — Kolloid. Zh. 27 (1965), 435.
232. Lipatova,T. E., I. S. Skorynina, and Yu. S. Lipatov. — In: Adhesion of Polymers, p.117.
      Moscow, Izd. AN SSSR, 1963. (Russian)
233. Kargin,V. A. and Yu. S. Lipatov. — Zh. Fiz. Khim. 32 (1958), 326.
234. Lipatov, Yu. S. and V.P. Privalko. — Vysokomol. Soed. A14 (1972), 38.
235. Wunderlich,B. — J. Phys. Chem. 64 (1960), 1052.
236. Lipatov,Yu. S. — DAN SSSR, 143 (1962), 1142.
237. Vonsyatskii,V. A., E. P. Mamunya, and Yu. S. Lipatov. — DAN UkrSSR B4 (1971), 330.
238. Moisya,E.G. and Yu. P. Egorov. — Teor. Eksp. Khim. 3 (1967), 131.
239. Egorov,Yu. P., E. G. Moisya, and M. A. Ar'ev.— Teor. Eksp. Khim. 3 (1967), 772.
240. Kuksin,A.P., L. M. Sergeeva, Yu. S. Lipatov, and L. I. Bezruk. — Vysokomol. Soed. A12
      (1970), 2332.
241. Rubin,R. — J. Chem. Phys. 51 (1969), 4681.
242. Shreiner,S. A. and P. I. Zubov. — Kolloid. Zh. 22 (1960), 727.
243. Zubov,P. I. and A. G. Sanzharovskii. — Lakokras. Mat. 2 (1963), 48.
244. Azori,M., N. A. Plate,et al. — Vysokomol. Soed. 8 (1966), 759.
245. Kargin,V. A., T. A. Sogolova, and I. I. Kurbanova. — DAN SSSR 162 (1965), 1095.
246. Lipatov,Yu. S. and L. M. Sergeeva. — Vysokomol. Soed. 8 (1966), 1895.
247. Lipatov,Yu. S. — DAN SSSR 5 (1961), 69.
248. Bonstra,B. and E. Dannenberg. —Rubber and Plast. Age 82 (1958), 838.
249. Lorenz,O. and G. Parks. — J. Polymer Sci. 50 (1961), 299.
250. Kraus,G. — J. Appl. Polymer Sci. 7 (1963), 861.
251. Lipatova,T.E. and V. I. Ivashchenko. — In: Synthesis and Physical Chemistry of Polymers,
      Vol. 6, p.73. Kiev, "Naukova Dumka," 1970. (Russian)
252. Kargin,V. A. and V. A. Kabanov. — ZhVKhO 9 (1964), 6021.
253. Sergeeva,L.M., Yu. S. Lipatov, and P. I. Bin'kevich. — In: Synthesis and Physical Chemistry
      of Polyurethans, p.131. Kiev, "Naukova Dumka," 1967. (Russian)
254. Lipatov,Yu. S., L. M. Sergeeva, et al. — Vysokomol. Soed. B10 (1968), 816.
255. Trostyanskaya,E.B. — In: Fillers of Polymeric Materials, p.3. Izd. Doma Nauchno-Tekhn.
      Propagandy im. Dzerzhinskogo, Moscow, 1969. (Russian)
256. Kargin,V. A., T. I. Sogolova, P. Ya. Rapoport- Molodtsova. — Vysokomol. Soed. 4
      (1964), 2090.
257. Malinskii,Yu. M. et al. — Vysokomol. Soed. A10 (1968), 786.
258. Malinskii,Yu. M. et al. — DAN SSSR 160 (1965), 1128.
259. Lipatov,Yu. S., S. S. Krafchik, and Yu. Yu. Kercha. — DAN UkrSSR 2 (1971), 143.
260. Fabulyak,F.G. and Yu. S. Lipatov. — Vysokomol. Soed. B12 (1970), 871.
261. Lipatov,Yu. S., F. G. Fabulyak, et al. — Vysokomol. Soed. A13, No. 11 (1971), 2601.

SUBJECT INDEX